Engineering management of water quality

McGraw-Hill Series
in Sanitary Science
and Water Resources Engineering

Rolf Eliassen and Ray K. Linsley   *Consulting Editors*

Babbitt, Doland, and Cleasby   *Water Supply Engineering*
Eckenfelder   *Industrial Water Pollution Control*
Ehlers and Steel   *Municipal and Rural Sanitation*
McGauhey   *Engineering Management of Water Quality*
McKinney   *Microbiology for Sanitary Engineers*
Linsley and Franzini   *Water-resources Engineering*
Sawyer and McCarty   *Chemistry for Sanitary Engineers*
Steel   *Water Supply and Sewerage*

ENGINEERING MANAGEMENT OF WATER QUALITY

P. H. McGauhey
University of California
Berkeley

McGraw-Hill Book Company
New York   San Francisco   St. Louis
Toronto   London   Sydney

Engineering management of water quality

Library of Congress catalog card number: 67-28085
234567890 MAMM 754321069

44975

PREFACE

In organizing and developing this book the author intended to present
and interpret the subject of water quality management in such a manner
as to stimulate thinking on the part of the reader. Although its use as
lecture material originally motivated the writing, the author speaks to
a wider audience than students in formal classes. The book is useful both
in and out of the classroom, and it avoids the impersonal, formal approach
without sacrificing rigor or completeness.

When used as text material, the author believes that this work will
help the sanitary engineer develop an understanding of the public policy
framework, and the overall system in which engineered systems for upgrading
the quality of water make sense and which may be expected to function
harmoniously for quality objectives. It lays the groundwork for the more
conventional technical courses in the principles, theory, and design of
water supply and waste-water treatment plants. For the student of water
resources engineering, it is intended to develop an understanding of
the place of water quality control in water resources management, but
anticipates that the student will follow it with courses in the economic and
planning aspects of water resources development and in its philosophical,
institutional, political, and legal aspects as well.

For the practicing engineer and other readers, the work directs
attention to recent changes in the rationale underlying public policy in
the resources field, and to the broad systems concepts which modern
engineering must apply to public works, with enough review of current
technology and future possibilities to provide a basis for individual
evaluation of man's social goals and his present technological ability to
achieve them in relation to water resources.

This work on the water quality management aspect of the subject
of water resources development and management was originally prepared
by the author as a text for one of four graduate courses comprising a

Summer Institute in Water Resources for College Teachers of Engineering and Earth Sciences. This institute was conducted by the Utah State University on its campus at Logan, Utah, in 1965.

In a less complete form it was reproduced as one volume of the four-volume *Proceedings of the Institute* for that year. The institute proved to be so successful that it was held again in 1966. In anticipation of this second year, and for use in a new Water Resources Engineering Educational Series developed in early 1966 by the Engineering Extension Division of the College of Engineering, University of California, Berkeley, the context was revised and enlarged. The work was further refined by the author in the fall of 1966 for use as a text in the first of a series of graduate courses in sanitary and water resources engineering at the University of California, Berkeley.

The author is especially appreciative of the contributions of two of his associates, Dr. W. J. Oswald and Dr. R. E. Selleck, who prepared Chapters 17 and 19, respectively. He is indebted to the many writers and sources referred to in the text for permission to use their published data. He is especially grateful to Dean D. F. Peterson, Dr. Jay M. Bagley, and others of the staff of the Utah State University, for providing the stimulus and encouragement needed to bridge the gap between actual production and the chronic endemic intent to write "someday" which the author shares with many of his colleagues in the engineering profession.

P. H. McGauhey

CONTENTS

Engineering management of water quality

CHAPTER 1

Introduction:
Quality as a dimension of water

The idea that "quality" is a dimension of water that requires measurement in precise numbers is of quite recent origin. Ancient British common law, from which our concept of riparian rights derived, was content to state that the user of water was not entitled to diminish it in quality. But the question of what constituted quality was neither posed nor answered. Since quantity, likewise, was not to be diminished, it may be presumed that the lawgiver had a limited concept of water use such as that of turning waterwheels; or perhaps he anticipated that the patent absurdity of the law would relieve both user and barrister of any need to take it seriously. At any rate a precise definition of water quality lay a long way in the future.

More than half a century ago a Mississippi jurist said, "It is not necessary to weigh with care the testimony of experts—any common mortal knows when water is fit to drink." Today we find it necessary to enquire of both common mortal and water expert just how it is that we *know* when water is fit for drinking. Moreover, in the intervening years, interest in the "fitness" of water has gone beyond the health factor and we are forced to decide upon its suitability for a whole spectrum of beneficial use involving psychological and social, as well as physiological, goals.

Looking back on the history of water resource development, one is impressed that under pioneer conditions it was usually sufficient to define water quality in qualitative terms, generally as gross absolutes. In such a climate, terms such as swampwater, bilgewater, stumpwater, blackwater, sweetwater, etc., produced by a free combination of words in the English language, all conveyed meaning to the citizen going about his daily life. "Fresh" as contrasted with "salt" water was a common differentiation arising from both ignorance and a limited need to dispel it. If a ground or surface water was fresh, as measured by the human

1

senses rather than the analytical techniques of chemists and biologists, it little occurred to the user that it was any different than rainfall in producing crops. The Lord has the sun at his disposal whereas the farmer has to rely upon ingenuity and hard labor to deliver the water; but delivery rather than content of the delivered product occupied attention. Thus the pioneer in search of water for his agricultural needs was content with a crude definition of its quality.

Within such a conceptual framework it is understandable that in the early days of America men were motivated to appropriate a portion of the region's freshwater resource; to hold it through the establishment of "water rights," public policy, or other property ownership device; and to defend it against all others by every possible means. *Quantity* was the dimension of water to which men sought title. Definition of its quality was not necessary as long as the water was fresh. There was, of course, one exception to this general rule—water for human consumption. Definite quality goals were established in the interests of the public health, and protection of public water supplies became an objective of water management. However, only about 7 percent of the water used by man was ever involved in public water supply, hence there resulted surprisingly little understanding of quality as a major aspect of water which required evaluation. "Acre-feet" tended to define much of the beneficial use of water, and "rights" to such units, the major goal of thirsty man. In this connection it is interesting to note that in the semiarid West, notably during California's gold rush days, the miner was the farmer's chief competitor for water. Since the miner's demands for water quality were even more rudimentary than the homesteader's, his contribution to a clarification of the concept of water quality was nil. He did, however, modify the ancient concept of "riparian rights" by adding the "first in use" amendment—albeit at the point of a gun.

The result of the historical preoccupation with the quantity of water was a whole fabric of tradition, law, judicial decision, and gunfire which established a property right in water without modifying the ancient gross concepts of its quality. Realization that this property right had a missing and critical dimension came about as a result of a burgeoning urban-industrial-agricultural economy which began to force repeated use upon a relatively constant water resource as an answer to its demand for water. Leaving to subsequent chapters the task of giving scale to this disparity between resource and demand, suffice it to say here that as the percentage of the overall water resource represented by "secondhand" water increases, so does the need for attention to water quality and the parameters which describe it.

As the author has noted elsewhere [1, 2]:

A need to quantitate, or give numerical values to, the dimension of water known as "quality" derives from almost every aspect of modern industrialized society. For the sake of man's health we require by law that his water supply be "pure, wholesome, and potable." The productivity and variety of modern scientific agriculture requires that the sensitivity of hundreds of plants to dissolved minerals in water be known and either water quality or nature of crop controlled accordingly. The quantity of irrigation water to be supplied to a soil varies with its dissolved solids content, as does the usefulness of irrigation drainage waters. Textiles, paper, brewing, and dozens of other industries using water each have their own peculiar water quality needs. Aquatic life and human recreation have limits of acceptable quality. In many instances water is one of the raw materials the quality of which must be precisely known and controlled.

With these myriad activities . . . going on simultaneously and intensively, each drawing upon a common water resource and returning its waste waters to the common pool, it is evident to even the most casual observer that water quality must be identifiable and capable of alteration in quantitative terms if the word is to have any meaning or be of any practical use.

Thus it is that those unwilling to go along with the Mississippi jurist must express quality in numerical terms.

The identification of quality is not in itself an easy task, even in the area of public health where efforts have been most persistent. For example, the great waterborne plagues that swept London in the middle of the nineteenth century pointed up water quality as the culprit; yet it was another quarter of a century before the germ theory of disease was verified, and more than half a century before the water quality requirements to meet it were expressed in numbers. Even in 1904, when our Mississippi jurist spoke, children still died of "summer sickness" (typhoid) often ascribed to such things as eating cherries and drinking milk at the same meal; and scarcely a family escaped the loss of one of its members by typhoid fever. Yet when it came to defining the water quality needed to avoid this, the best we could do was to place on some of the "fellow travelers" of the typhoid organism numerical limits below which the probability of contracting the disease was acceptably small. Nor has this dilemma been overcome. In 1965, an outbreak of intestinal disease at Riverside, California, which afflicted more than 20,000 people and caused several deaths, was traced to a newcomer (*Salmonella typhimurium*) in a water known to be safe by "experts" watching the coliform index. So once again the search begins for a suitable description of quality.

A second dilemma which survived the struggles that codified and institutionalized our concepts of water quantity lies in the definition of the word "quality." While the dictionary may suggest that quality implies some sort of positive attribute or virtue in water, the fact remains that one water's virtue is another's vice. For example, a water too rich in nutrients for discharge to a lake may be highly welcome in irrigation; and pure distilled water would be a pollutant to the aquatic life of a saline estuary. Thus, after all the impurities in water have been cataloged and quantified by the analyst, their significance can be interpreted in reference to quality only relative to the needs or tolerances of each beneficial area to which the water is to be put.

Shakespeare has said, "The quality of mercy is not strained . . . ." And indeed it is not as long as mercy is defined in qualitative terms. One can but imagine the problems which might arise if it were required that justice be tempered with 1.16 quanta of mercy in one case and 100 quanta in another. Yet this is precisely what confronts us in establishing measures of the dimension of quality of water.

REFERENCES

1 P. H. McGauhey: "Folklore in Water Quality Standards," *Civil Eng., N.Y.,* 35(6), June, 1965.

2 P. H. McGauhey: "Quality—Water's Fourth Dimension," *Proceedings of the Water Quality Conference,* University of California, Davis, January 11-13, 1961.

CHAPTER 2

Some fundamental concepts

INTRODUCTION

In Chapter 1 it was pointed out that whereas tradition and legal actions
to institutionalize tradition have established a property right in water
in terms of gallons or acre-feet, no such rights to quality have evolved. The
principal reason cited is that for a long time man got along more or
less satisfactorily with general rather than specific characterizations of
water quality. There were, of course, important exceptions in the area of
public health, once the relationship between water and disease was
understood, but in general, concern for quality could be met by alternatives
other than management of quality itself. The custom developed of
using water only once and then discarding it; and if the result of this
practice was intolerable pollution of water supplies, it was always possible
to go further, beyond the populated community, for a source of clean
water. The Romans, for example, built their first aqueduct—the Appia—in
313 B.C., not because the bountiful springs were drying up or inadequate
in quantity, but because the ground disposal of filth for generations
had finally polluted the groundwater beyond the tolerance of the
aesthetic sense of Romans.

While the Roman experience reveals that water quality has been
of concern to mankind for a long time, alternatives other than applying
precise numbers to the attribute of water called "quality" for the purpose
of quality management were available until quite recent years. During
a long interval, therefore, there developed a formidable list of attitudes,
adaptations, dislocations, and uncoordinated actions which constitute
the problem of water quality management, to which subsequent
chapters are directed.

As a background to a full understanding of the interchange of
quality factors and of the principles which can be harnessed to quality
management goals through engineered systems, it may be useful to dwell
briefly on a number of fundamental concepts.

5

CONCEPT OF ENERGY
IN BIODEGRADATION

The biochemical instability of organic matter, particularly human sewage, is the phenomenon which has been of most concern to those charged with responsibility for managing or controlling the quality of water. In a manner discussed in subsequent chapters, it is involved in the natural self-purification of flowing water and in the quality changes of stored or ponded water. It is a major aspect of the cycles of growth and decay on which aquatic life—and indeed, all life—depends. And its management under optimum conditions is the objective of all engineered systems for treating or disposing of organic matter or of other waste which organic life might reduce.

Figure 2-1 illustrates the sequence of energy releases involved in the exploitation of organic matter by living creatures. When man (or higher animals) consumes food, he is concerned physiologically with two things: obtaining energy and repairing worn-out cells. Energy is obtained by burning up some of the carbon and by unlocking the bonds in molecules such as those of protein. Small amounts of nitrogen, phosphorus, and other elements are assimilated in new cells, but the worn-out cells are discharged with bodily wastes, albeit at a lower energy level. The result is that for

Figure 2-1   *Concept of energy loss by biodegradation.*

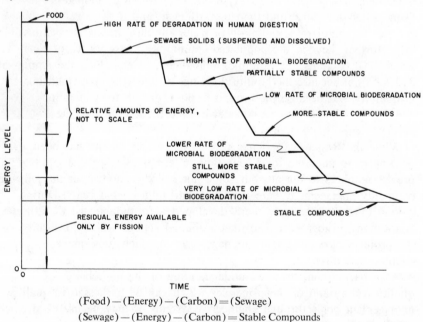

(Food) — (Energy) — (Carbon) = (Sewage)
(Sewage) — (Energy) — (Carbon) = Stable Compounds
(Stable Compounds) + (Solar Energy) + (Carbon) = Food

adult animals and humans the amount of these elements leaving the body each day is as great as the intake of such elements. The simple truth, then, is that although we may eat a variety of food for various reasons, about all our digestive systems extract is carbon and some of the bound energy. The remainder of our food passes out as a waste containing the original amount of all elements, less some of the original carbon.

As indicated in the figure, the energy content of the original proteins and other substances is still relatively high after human digestion. Therefore other organisms can get energy from further degrading it. These organisms are primarily the bacteria, each group of which in turn lives on the residue, or sewage, of a higher or preceding group, until finally what is left is a material of such low energy content that almost no organism can derive energy from further changing the chemical structure of the molecules. This residual material is then said to be "stabilized," and the process by which this situation was brought about is known as digestion, biodegradation, fermentation, etc., depending upon the circumstances surrounding the process. But in any case it is biodegradation that has occurred. "Bio" refers to the fact that chemical reactions have been brought about by biological agents (bacteria), while "degradation" refers to the lowering of energy contained in the substance by simplifying its molecular structure.

It is important to note, as illustrated in the figure, that as the energy level decreases, the residual energy becomes increasingly difficult to release. This introduces both a time factor and a microbial population limit, which become important in any engineered or natural system functioning at that particular level. It should be noted also that the number of steps in the degradation process, as well as the rates of energy conversion, may be vastly more numerous and complex than are illustrated in the figure.

The final stabilized product is by no means useless simply because bacteria are no longer attracted to it as a substrate. Rather, it is the fertilizer from which crops derive their nutrients. These nutrients, plus carbon dioxide from the atmosphere plus water, are all the plant needs in the way of building blocks. Through the agency of chlorophyll, these materials are locked together again with solar energy. The result this time is an increase in the energy level and the addition of carbon. So once again we have food from which we may burn carbon and extract energy.

As might be anticipated by the reader, waste-treatment processes are intended to provide engineered reactors to optimize the environmental conditions under which one or more of the steps in biodegradation takes place. Both engineered and natural systems deal with material at the energy level described in Figure 2-1 as "food," as well as with human wastes. Domestic waste waters invariably contain vegetable trimmings, meat scraps, and other organic matter discharged to the sewer in the process of preparing meals, cleansing dishes and clothing, and from various commercial and industrial activities. This material has not been degraded in energy by human digestion, hence it appears at a higher level of energy than do bodily wastes. From it bacteria can derive the energy that might have been ours had we eaten it. Hence a wider variety of bacterial species is involved in its reduction than in reducing human wastes. This same fact is abundantly true in the bio-degradation of the earth's vast amount of photosynthate and other dead organic matter in the cycle of growth and decay in nature.

CONCEPT OF AEROBIC AND ANAEROBIC CYCLES

The degradation of organic matter by a sequence of microbial populations can take place in two very important ways, both of which have significant quality implications in streams receiving domestic or industrial organic wastes. They can be depicted as cycles of growth and decay, such as are illustrated in Figures 2-2 and 2-3 for the elements of carbon, nitrogen, and sulfur. Similar cycles can be drawn for each of the many elements involved in organic compounds. They represent growth of living organic matter from nutrients, carbon dioxide, water vapor and solar energy, and its subsequent decay under two conditions of biodegradation:

1. By aerobic organisms under circumstances where free oxygen is available in suitable quantities.

2. By anaerobic organisms which must unlock from chemical compounds the oxygen needed to obtain energy from the oxidation of carbon.

It will be noted that the left-hand sectors of the two cycles are identical, since only the condition of biodegradation differs in the two cases. This sector indicates the principal manner in which stable compounds from the degradation of organic matter are incorporated into living plant matter at a higher energy level. Plant proteins, carbohydrates, etc., may become simply dead organic matter subject to biodegradation by bacteria, or may be consumed by humans and animals to be converted at a low conversion efficiency into a combination of living animal matter and bodily wastes. Ultimately both of these products become degradable dead organic matter, as indicated at the top of the cycle.

Decomposition of dead organic matter, including bodily wastes, in the presence of free oxygen is depicted in Figure 2-2. Here it may be noted that the initial products are unoxidized compounds of nitrogen and sulfur. These form the substrate of other organisms which carry the oxidation process further around the cycle. Ultimately, other organisms oxidize nitrogen and sulfur to the simple stable nitrate and sulfate forms. At each stage of the degradation, however, carbon dioxide is released by the organisms in oxidizing carbon to produce energy.

In contrast, the anaerobic decomposition depicted in Figure 2-3 produces initial and intermediate products that are unoxidized: Many of these are noxious and give the characteristic objectionable odor associated with putrifying sewage or animal flesh. Even the final products are unstable and hence subject to further oxidation at a later time when oxygen availability

Figure 2-2 *Nitrogen, carbon, and sulfur cycles in aerobic biodegradation.*

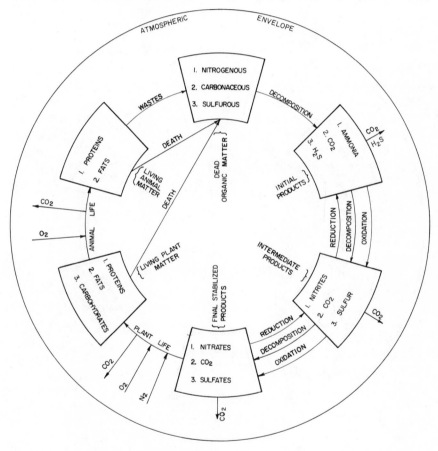

permits the development of aerobic organisms. Again it should be noted that at all stages carbon dioxide appears as the necessary product of energy production for the living organism. However, lacking sufficient oxygen to consume the carbon content of the waste, anaerobes get rid of the excess carbon by combining it with hydrogen. Since a great deal less carbon is oxidized in anaerobic decomposition than in aerobic, anaerobes release much less energy in the degradation process. In fact, the energy released under aerobic conditions is about thirty times that available to bacteria under anaerobic conditions. Consequently, the speed of aerobic degradation is much the greater, and there is an absence of odor in the process.

At the present stage of technology and economics, however, the use of both aerobic and anaerobic systems is unavoidable in the treatment of

Figure 2-3 *Nitrogen, carbon, and sulfur cycles in anaerobic biodegradation.*

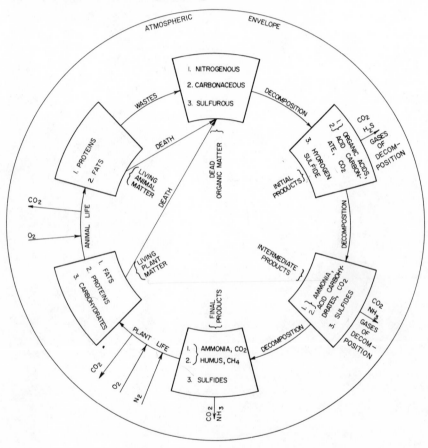

organic domestic and industrial wastes. Such wastes contain both dissolved and suspended solids, and we have as yet found no good way to make the nutrients in settleable particles available to a mass culture of aerobic bacteria. Hence the practice is to settle out the most readily separated suspended solids and confine them in a closed anaerobic system. Meanwhile the liquid portion, with its burden of unstable dissolved and finely divided particulate matter, is treated in an aerobic system.

CONCEPT OF BOD AND COD

By far the most widely deplored aspect of organic matter, notably human wastes, in water is the great quantity of oxygen required in aerobic degradation of the material in comparison with the capacity of water to dissolve free oxygen. The result, as discussed in detail in Chapter 18, is a depressing of the oxygen in water to levels inimical to aquatic life. Often this leads to anaerobic decomposition in a receiving water, with all its objectionable odors and other resource-destroying characteristics. As a measure of this oxygen-demanding property of a waste in water, a biochemical oxygen demand (BOD) test has been developed and is perhaps the best known measure of water quality in the world, in spite of its limitations of interpretation.

Within quite acceptable limits of definition it may be said that the "ultimate BOD" of an organic waste is the amount of oxygen required by microorganisms in carrying out the right-hand sector of the aerobic cycles of nitrogen, carbon, sulfur, phosphorus, etc., that is, the amount of oxygen needed by bacteria in reducing organic matter to stable compounds. This definition, of course, does not take into account the versatility and ingenuity of bacteria, a failure which has led to a vast waste of human energy in futile attempts to reduce BOD measurements to the precision of a simple pure acid-base reaction.

The oxygen demand curve

If a great number of samples of a water containing degradable organic matter, and diluted with water [1] containing adequate oxygen to maintain aerobic conditions throughout the degradation process, were placed in a controlled temperature incubator and a statistically significant number were examined each day for oxygen depletion, the plotted mean results would appear somewhat as shown in Figure 2-4. Analysis of such a typical oxygen demand curve reveals that BOD is exerted in two distinct stages:

1. A typical unimolecular reaction curve during the period when energy is being derived from oxidation of readily available carbohydrates and the splitting of protein and other molecules with the release of ammonia.
2. A somewhat straight-line curve during the "nitrification stage" when ammonia is being oxidized to nitrites and to nitrates.

Only the first of these two stages has been well explored experimentally and mathematically because:

1. The rate of oxygen demand during the first stage (10 to 20 days) is high, hence critical in receiving waters.
2. The oxygen demand during the second stage (3 to 6 months) is slower than the normal process of reaeration in nature, hence of secondary importance.

To utilize BOD as a practical yardstick of water quality the equation of the first-stage curve has been established [2] and typical deoxygenation constants evaluated experimentally for domestic sewage and a few other types of organic solids. Similarly, the effect of temperature on the BOD curve, such as illustrated in Figure 2-5, has been determined. Assuming

Figure 2-4   *Typical oxygen demand curve of aerobic decomposition of organic matter.*

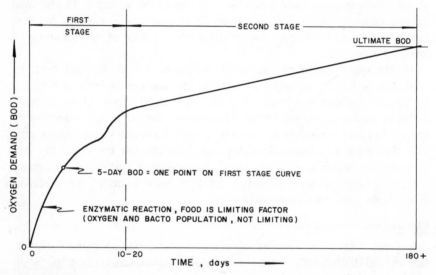

the validity of these values in any specific case, it is then necessary to obtain by experiment only a single point on the BOD curve in order to calculate the oxygen demand (BOD) to be expected at any temperature and time combination within the first stage of degradation. For a "standard" test, 5 days and 20°C have been adopted because:

1.   20°C is a typical summer water temperature, hence near the optimum for bacteria in the aquatic environment.

2.   At all temperatures likely to pertain outdoors, at which significant biological activity takes place, the first stage extends beyond 5 days. Furthermore, the statistical scatter of data at 5 days is acceptably low.

Although a discussion of the full significance and limitations of the BOD test is beyond the scope of this work, the importance of the BOD concept to water quality management cannot be overemphasized. When it is considered that a normal domestic sewage has a 5-day, 20°C BOD of 200 to 250 mg/liter, and that the value for industrial wastes may range from 3,000 mg/liter to more than a dozen times that figure, whereas a

13

stream fully saturated with oxygen at 20°C contains only 9.2 mg/liter of oxygen (20 percent less in saltwater), it is easy to anticipate the quick depletion of dissolved oxygen in any receiving water unless the dilution factor is quite large. It is likewise evident why the oxygen demand of decomposing organic matter has long been the most used parameter of water quality.

Figure 2-5 *Effect of temperature on oxygen demand. Slopes of second stage curves are greatly exaggerated.*

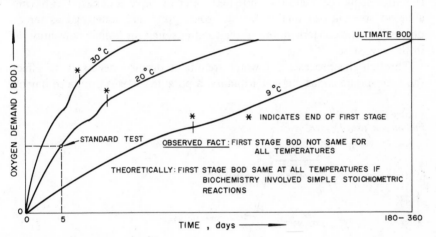

COD

The concept of chemical oxygen demand (COD) is that all organic compounds, with but few exceptions, can be oxidized to carbon dioxide and water. Therefore, a standard COD test has been devised [1] in which such an oxidation is carried out by a strong oxidizing agent to measure the pollutional potential of sewages and industrial wastes. In contrast with the BOD test, which measures only the biodegradable fraction, COD may measure toxic as well as biodegradable organic compounds. It is therefore applicable to many industrial wastes not readily analyzed for water quality factors by the sewage-oriented BOD test. Furthermore, the COD test is rapid, since statistical validity is not a function of time as in the BOD test.

CONCEPT OF BENEFICIAL USE

To the engineer concerned with water quality management, particularly with urban-industrial use, a concept of beneficial use which transcends the traditional is extremely important. Specifically, it should be clear that the collection, development, treatment, transportation, and distribution of a water supply, and its re-collection, retreatment, transportation, and return back to the water resource after it has become a "waste" water, are all facets of a single beneficial use of water. Obvious though this concept may seem, it is not characteristic of the background of water resource engineers. Formal educational courses normally separate the subjects of "water supply," "sewerage," and "industrial wastes"; and for reasons of the historical development of water quality concern, engineering practice has been similarly compartmentalized.

Figure 2-6 illustrates the quality changes during municipal use of water in a time sequence. Here the untreated water resource is depicted as having a broad spectrum of quality factors, some good and some bad as far as domestic use is concerned, but in the total unsuited for human consumption in the raw state.

Through the processes of water treatment deemed necessary to effect the appropriate change in water quality, a potable water is produced having

Figure 2-6 *Quality changes in domestic use of water.*

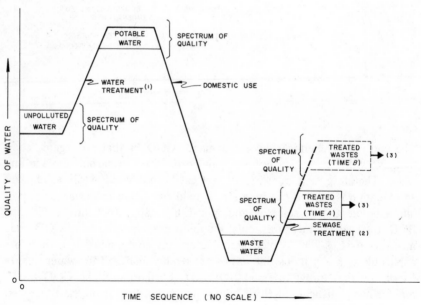

(1) Processes include sedimentation, chemical coagulation, filtration, and sterilization.

(2) Processes include sedimentation, activated sludge, trickling filters, chlorination, ponding and disinfection.

(3) Effluent returned to resource pool.

a different spectrum of quality factors. By the addition of a small amount of objectionable solids, domestic use degrades water quality to a low level. The quality changes necessary to upgrade this waste water then become a matter of concern of sewage treatment. In the actual case, the treatment process is carried only to the point demanded by regulatory agencies for protection of other beneficial uses. That is, there is seldom any attempt to carry to complete stabilization the biodegradation of organic matter. However, as noted in Chapter 18, stabilization itself may not be enough to make the treated water acceptable in the water resource.

The dashed lines in the figure represent an increase in treated wastewater quality as the rising standards of sanitation, or intensive demands on the water resource, call for greater application of technology. Ultimately, as the quality of treated waste water approaches that of the original natural resource, the concept of "water reclamation" or "water renovation" is generated.

CONCEPT OF TRANSPORT
SYSTEMS VERSUS SINKS

Fundamental to any practical approach to water quality management, or even to an understanding of the problems of water quality per se, is a concept of transport systems in contrast to "sinks" or reservoirs. Like all matter, the factors which constitute water quality are mutable but indestructible; hence, either in the unaltered or altered state they must be sequestered or tolerated somewhere in the air, the water, or the land resource. Quite aside from interference with other uses of the resource, these three elements differ vastly in their capacity to accept matter; and from this fact derives an important concept of sinks and transport systems.

From the viewpoint of geological time it may be said that the ocean is the ultimate sink into which all quality factors may be discharged—the single reservoir in which water may abandon its load of dissolved and suspended solids. Such a time scale envisions the erosion of the land and its slow migration down to the ocean. On a more limited time scale suited to man's reckoning, however, the land also may be considered as a sink, and we so use it in stockpiling solids which are either excluded from water or removed from it in the course of quality management. Thus in the context of vast capacity to accept insult, the oceans and the land may be considered as sinks.

In contrast with the land and the ocean, the atmosphere and certain sectors of the water resource have a very restricted capacity to accept or carry extraneous fractions. They are looked upon as transport systems because in reality they serve mainly to carry wastes from some point of origin to one or both of the sinks. The atmosphere, for example, although blanketing the entire earth to a depth of many miles, acts largely to pick up particulate matter from some point of discharge and scatter it over the land and sea. To a significant degree, gases are likewise transported to be removed by the vegetation on the land or absorbed in seawater. The important point is that in spite of its vast volume the atmosphere has a limited capacity to hold waste products. Physical limits, as regards solid particles, are somewhat analogous to the saturation capacity of water, i.e., its ability to dissolve only so much of a given compound. In its ability to accept gases such as CO_2 and the oxides of nitrogen and sulfur, the atmosphere is not limited by the same factors as those involved in particulates. Here it behaves more like a sink but with more immediate dangers to man, such as the "greenhouse effect" of CO_2 which might result in melting of the earth's ice caps and drowning out most of man's urban communities. In contrast, the buildup of salts in the ocean, although analogous to CO_2 buildup in the atmosphere, offers less immediate prospects of catastrophe for man.

In terms of water quality, the atmosphere as a transport system is most important in that it is the sole agency by which essentially pure water is distributed over the earth. Thus water quality management includes a concern for limiting the load of pollutants in the atmosphere to less than its actual physical capacity—a necessity that in reality is made academic by the much lesser threshold involved in sustaining life through breathing.

In a similar fashion, flowing water is a transport system rather than a sink. Its physical ability to accept and carry away quality factors is limited; and in the actual case, the maintenance of quality suited to human and aquatic life reduces the transport capacity to values well below the physical limits.

Lakes and ponds, estuaries, bays, and even the shallow ocean waters overlying the continental shelves likewise behave more as transport systems than as sinks in that they have limited capacity and tend to purge themselves when the addition of "pollutants" ceases. In contrast, the land and ocean tend to retain whatever is stored there. Quite obviously the concept of sinks and transport systems overlooks such important facts as their behavior as reactors and the many complex subtleties in the life cycle of each. Nevertheless, the concept is valid and extremely significant in practical water quality management.

REFERENCES

1 *Standard Methods for the Examination of Water and Waste Water,* 12th ed., American Public Health Association, 1965.

2 Clair N. Sawyer: *Chemistry for Sanitary Engineers,* 1st ed., McGraw-Hill Book Company, New York, 1960.

CHAPTER 3

Quality interchange systems

INTRODUCTION

The concepts set forth in Chapter 2 are fundamental to an understanding of the engineered systems designed to upgrade the quality of waste streams from individual beneficial uses to a level acceptable for discharge into the local receiving waters. Concern with such systems was the primary characteristic of engineering for water quality control as long as the concept of quality management was that of control of "pollution." In this context, undesirable quality factors in a waste water were viewed as "pollutants"; the discharger was a "polluter," and the remedy was "pollution control." Consequently the emphasis was upon the quality of waste discharges even though the goal of pollution control may have been that of maintaining the quality of the water resource.

In recent years, as the demand for water has approached in magnitude the available supply, there has been a profound expansion of the goals of quality control agencies. First, it was protection of the public health. Later, as both the amount and the variety of waste discharges grew, concern for the protection of the several other beneficial uses was expressed in legislation and in the regulations of newly created pollution control agencies. Recently, aesthetic or social goals of water quality have been added to the growing list of objectives of water quality management. The result is that the emphasis of management has become resource-oriented, in contrast with its former orientation to use. Concern has shifted from pollutants to the water resource itself. In this context resource protection becomes the object of public policy, and the various beneficial uses appear as subsystems which must be managed in a coordinated and integrated fashion to achieve resource policy objectives. While this does not mean that engineering is no longer concerned with the techniques and technology of changing the quality of water used and returned by cities, industries, and agriculture, it does mean that systems engineering

17

in the modern sense must now precede the traditional design of engineered systems in the management of a water resource.

For somewhat obvious reasons the first application of systems engineering to water resource management was in the control of quantity rather than quality of water. However, since quality is always one of the dimensions of water, no aspect of water can be managed without involving it. It is therefore the purpose of this chapter to present a concept of the systems with which the engineer involved in water quality management must one day be concerned. Although a full discussion of all these systems is beyond the scope of this work, some of them set the stage for subsequent chapters.

RISE OF A NEED
FOR SYSTEMS DESIGN

The network of streams which comprise an individual sector of the surface-water resource has long been referred to as a "river system," but more in a geographical sense than as an engineering concept. When it was isolated from adjacent sectors of the map by a line delineating a "drainage basin," the system took on the characteristic elm leaf appearance shown in Figure 3-1. Water resource development, however, was not early directed to the entire leaf. Instead, engineers were concerned with individual structures —a power dam on this fork of the river; on that one, a flood control works; on another, a diversion structure. Often this sort of development went on over a long period of time and under the guidance of a variety of agencies. The significance of such a piecemeal, unintegrated, limited-objective type of development from an engineering viewpoint is that there was no overriding goal of providing structures that would function in harmony with each other. Hence, while nature may have provided a river system, there was nothing particularly systematic about the way man developed it.

The end to this approach was at hand when it was recognized that what was going on was detrimental to use of the full potential of the water resource, and public policy moved in the direction of comprehensive, integrated development for multipurposes. Such a policy was not, however, based upon the availability of a systems engineering methodology, but it did create a situation in which such an approach was imperative if the physical installations involved were to be managed in such a manner as to achieve the objectives of multipurpose development. Here it is important to note that "multipurpose" applies in two different contexts in water resources systems. In the first, it refers to a number of beneficial uses within a single project, e.g., water supply, flood control, irrigation, navigation, fish and wildlife, recreation, and wastes dilution. In the second, it involves multipurpose objectives of public policy, e.g., protection of the public health, protection of beneficial uses, and aesthetics.

NATURE OF SYSTEMS
IN WATER RESOURCE
DEVELOPMENT

System of
multiple benefits

The best known system in the water resources development and management field is that depicted in Figure 3-1. Here a series of dams on tributaries

Figure 3-1 *Multipurpose river*
system development.

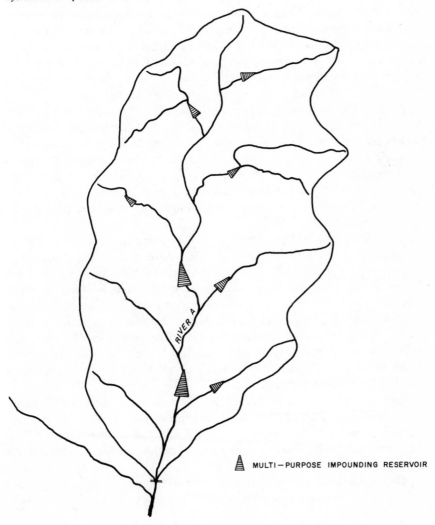

MULTI—PURPOSE IMPOUNDING RESERVOIR

and on the main stream itself must be managed as a system for a variety of objectives such as:

1. Routing of floods to minimize downstream disasters
2. Making flood-storage capacity available at the proper season
3. Maintaining optimum output of power
4. Augmenting low flows during dry seasons
5. Discharging water to beneficial uses on a predetermined flexible schedule

Examples of this type of system are at once evident in the Tennessee Valley Authority and in the developments in the Missouri River Basin. And indeed, flood routing via computers is an increasingly common practice. For example, Professors J. A. Harder and H. A. Einstein of the University of California at Berkeley constructed for the Corps of Engineers an analog of the Arkansas River above Kansas City [1] which can quickly trigger the necessary spillway control action to route a flood resulting from any amount of rainfall on any or all parts of the basin.

At first thought it might appear to the uninitiated observer that a system suited to objectives such as the five listed above is a relatively straightforward case for a mathematics-statistics approach. And, indeed, it is as close to such a system as occurs in the water resources management field. A mathematical model can be set up describing quantitative inputs statistically derived. Feedback from hydrological data, power demands, flow demands, etc., can then be used to correct the solutions to current reality, at least on a seasonal basis.

The fallacy in such logic, however, is the assumption that the model can be made to describe reality in a situation where both multiuse and multiobjectives of a qualitative nature exist. Specifically, draining the flood and power storage pools interferes with recreationists, makes boat landing facilities difficult to maintain, and leaves camps and cottages high and dry. The shallows habitat of fish and wildlife is catastrophically affected. Furthermore, encroachers on the original flood plain suffer damages by even, controlled floods at one season, and waste dischargers are short of dilution water at another. Finally, the holdback of sediments by dams may cause the loss of shoreline resources.

Each of the interests concerned has its own vocal supporters. In fact, several of these interests are listed as beneficiaries in the original multipurpose concept of the system. The feedback from all these sectors takes the form of public outcry, newspaper editorials, political opportunism, and ballots.

Thus qualitative inputs appear in even the simplest system of quantity management in water resources for multipurposes.

The natural
interchange system
of quantity and quality

As a baseline for management of engineered systems, particularly in the
selection of process and in the establishment of criteria for subsystems
design and evaluation, it is necessary to consider the interchange system
that exists under natural conditions. Such a system is depicted in Figures 3-2

Figure 3-2 *Natural interchange*
system. The percent sign (%)
indicates a fraction of the average
United States rainfall of 30 in. [2]

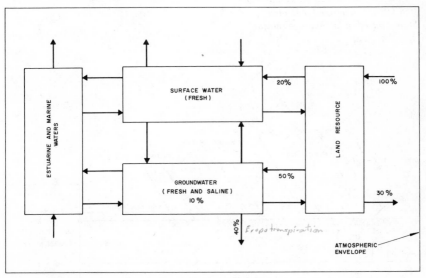

and 3-3. The figures show the paths by which water quantities and quality
factors are interchanged. Disposition of the average 30 in. of annual rainfall
upon the land surface of the United States is noted on Figure 3-2 as per-
centages. They show that of the rain that falls upon the land in the United
States, 30 percent is reevaporated, 20 percent runs off into ponds, streams,
and lakes, and 50 percent enters the ground. Of this latter, four-fifths, or
40 percent, of rainfall leaves by evapotranspiration. The remainder appears
as the groundwater fraction of the freshwater resource, overflowing into
fresh or saline surface waters when the storage capacity of the ground is
filled.

Man-modified
interchange systems
of quantity and quality

Although Figures 3-2 and 3-3 describe the system of quality and quantity interchanges in an unmodified situation, their usefulness is largely to identify the points where engineered subsystems may alter the natural pattern, and where the quality factors with which such subsystems are to be concerned must be identified. The major problem of engineering develops around the man-modified natural system by which water resources are put to beneficial use.

Figure 3-3 *Quantity interchange for meteorological input. The percent sign (%) = average for United States, based on an average rainfall of 30 in./year.*

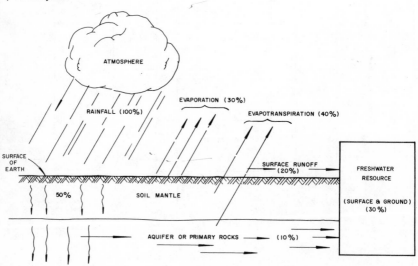

Figure 3-4 shows schematically the interchange system for major beneficial uses which depend upon withdrawal of water from the freshwater resource and on the discharge of return flows back into the resource. Considering the combined surface and ground water as a freshwater resource pool, the figure identifies the bridges across which water is withdrawn and returned either to the resource or to the atmosphere. Thus, it identifies the points where quality factors must be identified and where engineered subsystems can be introduced to change the quality of water to that suitable for beneficial use, or to make it suitable for return to the resource pool pursuant to both public policy and the quality needs of others. Similarly, it identifies the points where subsystems may recycle water for quality conservation.

The system shown in Figure 3-4 represents the type of system which must be reckoned with *in toto* if engineers are to achieve the objective of maintaining a quality of the resource in line with the needs of uses which do not depend upon withdrawals, e.g., recreation and wildlife, and with the aesthetic goals currently in vogue, such as clean water per se, quality of environment, and quality of life itself.

Figure 3-4 *Man-modified quality interchange system.*

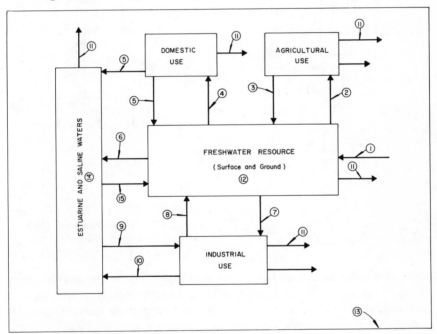

Local resource-need considerations

The concept utilized in Figure 3-4 of the freshwater resource as a single resource pool is useful at certain levels of analysis of the system. However, the acute problems of water quantity and quality exist where the people are; and people are concentrated in urban centers rather than being uniformly distributed on the land. The water resource is likewise concentrated in certain areas, but by quite different laws than govern the assemblage of men. For example, in California [3] 70 percent of the water is in the northern one-third of the state and 77 percent of the man-generated need is in the southern two-thirds. Engineers therefore find it necessary to recast the system shown in Figure 3-4 into a series of local (areal) systems, two of the most obvious of which are represented graphically in Figure 3-5.

Case 1 in Figure 3-5 is a relatively simple situation in which no very serious pressure exists to relate ground- and surface-water withdrawals to competing needs. In this case the systems analyst, whether dealing with the

Figure 3-5   *Local resource-need relationships.*

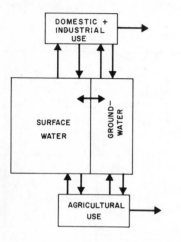

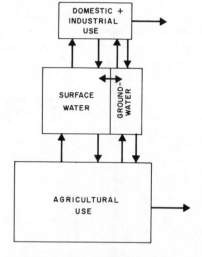

CASE I   LOCAL RESOURCE GREATER
THAN LOCAL NEED

CASE 2   LOCAL RESOURCE LESS
THAN LOCAL NEED

overall situation or the subsystems for managing quantity and quality, might successfully detour around some of the nonmathematical inputs, such as the difference in attitude of a beneficial user concerning the quality of his influent and return flows, in reaching engineering design goals.

Case 2, however, is in itself unsolvable. Such a system presents only two major alternatives:

1. Combine one or more case 2 and case 1 systems into a super system having the characteristics of case 1, or at least its lower limits.

2. Confine the water use in case 2 to the potential of the resource via a system of allocations, recycling, and reclamation of water.

The first of these alternatives is similar to the situation discussed in relation to Figure 3-4—redistribution of quantity of water—except that quality factors, provincialism, water rights laws, vested interests, and the general contrariness of mankind become inputs; and the feedback, being unresponsive to statistical methods, alters the mathematical solution in unpredictable ways. Nevertheless, the situation is real and a systems design is incumbent upon the engineer.

The second alternative introduces a public policy factor of economics and is worthy of separate consideration.

Economic optimization

If the system indicated as case 2 in Figure 3-5 is to be considered without relevance to other local or regional situations, the problem becomes one of allocation of a limited resource to various beneficial uses in accord with public policy. However, the onset of the need for such a decision has been so precipitous that generally no suitable public policy exists. Short-range political solutions based on pioneer objectives of land occupancy, homesteading the West, etc., no longer offer solutions; hence it becomes necessary to analyze the system in order to establish a basis for public policy.

Almost automatically the search then takes the form of examining the economic consequences of various alternatives, based on a whole series of assumptions concerning the relative stability of all sectors of the world's economy and the stability of the system itself. Without much help from the computer we can establish the criterion that the most logical policy objective in this case is that of producing, by the allocation of water, an optimum growth of the economy.

To such an end, studies by engineers and economists are under way to devise methods of programming such an optimizing system. An approach which might be described as "carefree mathematics" is evolving. It seeks to apply linear programming to nonlinear systems, sometimes subject to qualitative inputs, by a combination of mathematical ingenuity and the introduction of value judgments when necessary into a dynamic model. The problem is a real one in multipurpose water resources management, and neither the complexity of the system nor the impurity of the mathematics with which it must be approached at the present time relieves the engineer and his coworkers of the necessity for dealing with it.

Quality evaluation

The advance in demand of beneficial uses on a relatively constant water resource has not been paralleled by an increase in the effectiveness of engineered subsystems designed for quality control. The result is both a pressure for more costly water treatment and an overall decline in the quality of the resource. This has given rise to the concept of "quality" as a dimension of water which should be subject to numerical description and to economic evaluation.

Figure 3-6 represents a system in which streamflow regulation for low flow augmentation is possible, perhaps at the expense of other interests in a multipurpose reservoir. Domestic and industrial wastes are introduced which may be varied in volume-to-streamflow ratio, in composition, and in degree of treatment. Dilution of the stream by tributaries carrying variable volume and waste-water inputs is added to the model. Self-purification of the stream appears as a quality-upgrading factor; and the stream is subject to a spectrum of possible regulatory "standards." Finally, a second community withdraws water for municipal uses and may apply various degrees of treatment.

Despite the obvious complexity of the model, there is a real need to harness the techniques of systems analysis to a determination of what fraction of the program of water quality restoration should be ascribed to each natural and engineered system (given any set of discharge values,

Figure 3-6 *Typical model for quality evaluation study.*

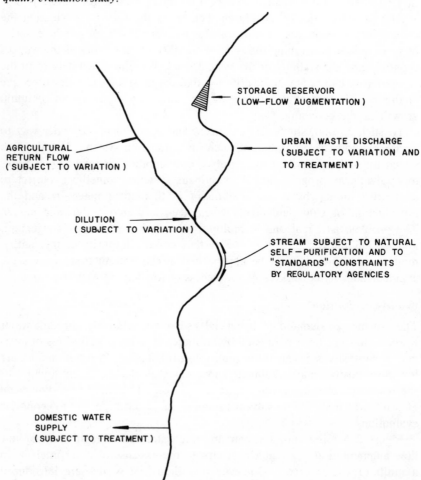

flows, regulatory constraints, and public policy) in order to achieve public water quality policy objectives at the least cost. Similarly, it is important to determine the value of water quality to each beneficial use when both quantity and quality may be the variables.

Although work on the economic evaluation of water quality is only beginning and the task is a formidable one, there is no escaping the fact that rather suddenly there has arisen a need to manage both quality and quantity of the water resource of the nation if the inevitable return flows are not seriously to constrain man's freedom to use that resource as he wills. Methods of management will have to evolve from an ability to deal with the entire systems of which Figure 3-6 is but a modest example.

Quality relationships in water-air-land resources

The final and most difficult system confronting the engineer concerned with multipurpose water resource development superimposes a multiresource relationship upon multiplicity of beneficial uses and multiplicity of policy objectives. Figure 3-7 depicts the vastly involved system in the man-modi-

Figure 3-7 *Regional interresource quality interchange system.*

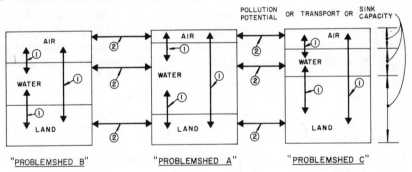

① Interresource interchange
② Interarea interchange (boundary condition)

fied world where the quality of land and air resources cannot be divorced entirely from the management of water. Here again the problem exists where people are concentrated. Thus a system involving water-air-land interchanges and divided into separate "problemsheds" is becoming a recognized situation. Management of wastes within one problemshed or subsystem then becomes dependent not only upon the relative magnitude of the air, water, and land resources in the immediate area, but also upon the boundary conditions, or flux, between a number of such areas.

The figure is intended to convey the concept of interresource interchange of quality factors in locales having widely different combinations of resource potentials, yet interdependent in the face of overall public policy on quality of the environment. Although no progress has yet been made on analyzing such a system for design purposes, the rationale, the public policy, and the concentration of quality-depressing factors are rapidly emerging and an early consideration of such a system is inescapable.

SUMMARY

Almost overnight the piecemeal unintegrated engineering approach based on building a structure here and there became hopelessly inadequate; and the engineer fitted to its requirements, a mere technician. Whole systems of interregional and interresource interchange of quality and quantity have now to be understood and analyzed in the process of engineering design. Moreover, multiple use and multiple policy objectives introduce qualitative inputs and generate nonmathematical and nonstatistical feedbacks. The first goal of systems analysis, therefore, must often be the presentation of alternatives from which political decision can lead to public policy. Policy considerations must then be fed back into the system in the production of a set of technical and economic alternatives for further policy decision. Design of the essential subsystems, which once constituted the whole of engineering design, now appears only as a final step. This is not to gainsay that without this step engineering works would not exist, but rather to stress again the compelling nature of systems engineering which now precedes it.

REFERENCES

1   James A. Harder: "Analog Models for Flood Control Systems," *Proc. ASCE*, 88(HY-Z), March, 1962.

2   George E. Symons: "The Hydrologic Cycle," *Water Sewage Works*, 100(7), July, 1953.

3   Harvey O. Banks: "Water . . . Today and Tomorrow," Brochure, California State Department of Water Resources, 1959.

CHAPTER 4

Quality changes by domestic use

INTRODUCTION

The man-modified system of quality interchanges depicted in Figure 3-4 identifies the various bridges of withdrawal and return of water at which quality management is feasible. Considering that in a primitive or "wilderness" state the quality of water is generally suited to a variety of beneficial uses, it seems logical to leave for later discussion the changes in that quality which occur in the natural course of events, and to examine first the nature of the various separate uses that draw upon the resource pool and return effluents to it. Domestic water supply, being the most necessary for the continuance of human life, is generally considered the highest beneficial use of water. Similarly, the nature of its return waters has historically caused most concern to man for health and aesthetic reasons. Therefore, municipal use seems worthy of first attention; and for reasons of the difference in quality management procedures in the two cases, domestic and industrial use are considered separately even though both may draw upon the same municipal supply and discharge return flows through a common sewerage system.

Domestic water use is normally considered unconsumptive. However, lawn sprinkling, street flushing, evaporative cooling, energy production, and similar activities may account for as much as one-third of domestic use. Thus a water demand of 150 gallons per capita per day (gpcd) may result in a waste-water return flow of only 100 gpcd. The principal task of this return flow is to transport wastes resulting from the life processes of man and of the activities attendant to preparing and managing his food. Although as indicated in Chapter 2 the biochemical instability of the organic matter in these wastes may make them extremely objectionable, it is hard to justify the dedication of some 2,000 tons of water to transport a single ton of solids which have survived waste-treatment processes. Certainly it is without parallel in the history of transportation to send so vast a train to carry so small a load. And equally certain is the need for quality management to include concern with transportation systems. The purpose of this chapter, however, is to consider the nature of the transported load in terms of the quality factors it returns to the freshwater resource pool.

GENERAL CHARACTERISTICS
OF DOMESTIC RETURN WATERS

Some quality factors in human wastes are customarily described in general terms; others are expressed in specific numbers. Both of these evaluation techniques are important in one way or another in the development and operation of engineered systems for controlling or managing the quality of domestic return flows. Three terms are used to describe sewage generally. All are of significance in sewage treatment although not all are equally rigid in definition. They are: condition, concentration, and composition.

Condition refers to the age of sewage. Three categories of condition are usually identified and defined:

1. *Fresh* Sewage in which the dissolved oxygen concentration is not materially less than that of the municipal water supply which went into it. In terms of the concept of biodegradation, a fresh sewage is one in which the oxygen-demanding processes of decomposition have not yet become evident.

2. *Stale* Sewage in which the dissolved oxygen content has been depleted to near zero by biological degradation.

3. *Septic* Sewage in which biodegradation has set in, and a population of organisms in scale with the food supply has been established.

The condition of a sewage in a sewer line may govern the measures necessary to protect the sewer and other concrete structures from hydrogen sulfide damage. It may also govern the nature of pretreatment (such as preaeration) in the waste-treatment plant itself.

Concentration refers to the strength of a sewage, usually measured by its 5-day, 20°C BOD, although it may be measured in terms of relative amounts of volatile (biodegradable) solids. Commonly used categories are:

1. *Weak* Sewage in which the BOD is below about 180 mg/liter.

2. *Average or medium* Sewage in which the BOD is in the range of from 200 to 250 mg/liter.

3. *Strong* Sewage which has a BOD above 280 to 300 mg/liter.

In no case is the concentration of sewage a rigidly defined quality factor, nor is the boundary line between one category and another sharply identified. Nevertheless, the concentration of sewage is a useful concept in explaining the difference in behavior of a treatment process in different cases, and in the making of preliminary estimates of process costs and of type of system suited to the quality goals of treatment.

Composition of sewage refers to its content of various waste-contributed quality factors, oxygen-demanding potential, products of biodegradation, etc. Table 4-1 presents a typical analysis of strong, medium, and weak domestic sewage in the United States.

Table 4-1 *Typical composition of domestic sewage (All values in mg/liter)*

Constituent	Strong	Medium	Weak
Solids, total	1000	500	200
Volatile	700	350	120
Fixed	300	150	80
Suspended, total	500	300	100
Volatile	400	250	70
Fixed	100	50	30
Dissolved, total	500	200	100
Volatile	300	100	50
Fixed	200	100	50
BOD (5-day, 20°C)	300	200	100
Oxygen consumed	150	75	30
Dissolved oxygen	0	0	0
Nitrogen, total	86	50	25
Organic	35	20	10
Free ammonia	50	30	15
Nitrites (NO_2)	0.10	0.05	0
Nitrates (NO_3)	0.40	0.20	0.10
Chlorides	175	100	15
Alkalinity	200	100	50
Fats	40	20	0

SOURCE: Babbitt and Baumann [1]. Reproduced by permission of the publisher.

The quality factors listed in Table 4-1 are but part of the precise description of the composition of a sewage. A full chemical analysis would include also the results of analytical measurements of the various anions and cations occurring in the sample, many of which were present in the domestic water supply and identify its quality. Two other types of analyses are used to characterize domestic return water. These identify its physical and biological attributes. Results of the first are expressed in qualitative terms or in numerical values along an arbitrary "standard" scale. In the second, the results, though expressed numerically represent statistically probable values.

PHYSICAL ASPECTS OF DOMESTIC RETURN WATERS

Some physical characteristics of the return water from domestic use and their general significance in the water resource are summarized in Table 4-2.

Table 4-2 *Physical characteristics of domestic waste water*

Characteristic	Cause	Significance	Measurement
Temperature	Ambient air temperature. Hot water discharged into sewer from home or industry.	Influences rate of biological activity. Governs solubility of oxygen and other gases. Affects magnitude of density, viscosity, surface tension, etc.	Standard centigrade or Fahrenheit scale
Turbidity	Suspended matter such as sewage solids, silt, clay, finely divided organic matter of vegetable origin, algae, microscopic organisms.	Excludes light, thus reducing growth of oxygen-producing plants. Impairs aesthetic acceptability of water. May be detrimental to aquatic life.	Light scatter and absorption on an arbitrary standard scale [2]
Color	Dissolved matter such as organic extractives from leaves and other vegetation (tannins, glucosides, iron, etc.), industrial wastes.	Harmless generally, but impairs aesthetic quality of water.	Light absorption on a standard arbitrary scale [2]
Odor	Volatile substances, dissolved gases, often produced by decomposition of organic matter. In water it may result from the essential oils in microorganisms.	May indicate presence of decomposing sewage. Affects aesthetic quality of water. As a test of sewage it may serve, for example, as a guide to condition of sewage when it reaches the treatment plant.	Human sense of smell, qualitative scale, and concentration at threshold of odor

Taste	Materials producing odors. Dissolved matter and various ions.	Impairs aesthetic quality of water.	Not measured in unpotable water
Solid matter	Dissolved and suspended organic and inorganic solids.	Measures amount of organic solids, silts, etc., hence is a measure of the extent of sewage pollution or the concentration of a sewage.	By gravimetric analysis techniques [2] for the following: Total solids, total volatile solids, total fixed solids, suspended solids, and dissolved solids*

*Total solids: As a measure of the amount of total dissolved and suspended matter. Total volatile solids: As a measure of the decomposable organic matter. Total fixed solids: As a measure of inorganic grit and dissolved inorganic matter plus ash of organic matter. Suspended solids: As a measure of material which might be removed by settling. Volatile and fixed suspended solids: As a measure of the decomposable organic matter and of inorganic matter. Dissolved solids: As a measure of material to be removed by secondary sewage treatment processes. Volatile solids: As a measure of organic matter which may be decomposed (may exert BOD). Fixed solids: As a measure of residues which may add to the burden of the effluent (lower its quality).

BIOLOGICAL CHARACTERISTICS
OF DOMESTIC RETURN WATERS

Domestic return water contains vast numbers of bacteria—up to 20 million or more organisms per milliliter—originating in the wastes discharged from the human body and on other material introduced into the sewer. The feeding activities of some of these organisms result in the degradation of sewage as discussed in relation to the cycles of growth and decay of organic matter (Chapter 2). Some of the bacteria in human wastes are pathogens, but their number is small in comparison to the total microbial population. However, the number of pathogens necessary to produce disease is also quite small.

The variety and number of enteric pathogens in sewage depend upon the diseases endemic in the contributing population. Thus gastroenteritis may be most common in United States domestic waste water, whereas elsewhere typhoid, paratyphoid, or cholera might be equally likely. The same may be said of other biological agents such as the eggs of stomach and intestinal worms, cysts of amoeba, and the enteric viruses.

Most of the bacteria in domestic wastes are concerned with biodegradation of organic matter. Consequently they cause the composition of a sewage-polluted water to change continually as the process of stabilization proceeds. Thus the concentration of several quality factors in an untreated domestic return water is also subject to change with time.

Attempts to express the biological quality of domestic effluents in precise terms are normally confined to two types of examinations:

1. The total number of organisms per unit of water that will grow on a standard substrate during a specified time period under aerobic conditions, and

2. The most probable number of intestinal organisms of the coliform group present per unit of water and viable under standard test conditions.

Results of the first of these tests give some indication of the concentration of unstabilized organic solids at the time of the test. Interpretation of results of the second is based on the presumed validity of statistical relationships between coliform densities and the incidence of disease from intestinal pathogens at such coliform densities.

As will be discussed in Chapter 12, interpretation of statistical measures of biological quality of water has been practical only at the limit below which the health risk is an acceptable one. Above that limit the relation of risk to quality factor concentration is subjective only. Furthermore, only coliform organisms are normally measured. Efforts to measure the quality of domestic return waters by observation of other organisms, such as fecal coliforms and viruses, produce qualitative data which are as yet difficult to interpret.

In summary it may be said that the return waters from domestic use contain, or may contain, biological agents dangerous to human health but difficult to evaluate in terms of precise measures of water quality. This, as later discussed, imposes a problem in water quality management at both a policy and a technological level.

STATUS AND QUALITY OF DOMESTIC RETURN WATERS

In order to assess the effect of domestic use of water upon the freshwater resource pool, and hence its importance as a factor in water quality management, it is necessary to know something of the magnitude of such use and of the status of quality management at this point in the interchange system.

From the preceding discussion of the characteristics of domestic wastes it may be concluded that the return water from municipal use carries organic matter typical of that found in the normal cycle of organic growth and decay in nature. It reaches the water resource pool at various levels of energy residual depending upon the degree of treatment or length of time it has been undergoing biodegradation. It carries oxygen-demanding unstable organics on which bacteria feed. In addition it carries microbial agents which may be pathogenic.

Table 4-3 gives a rough idea of the status of the organic matter in domestic return water subjected to various treatments:

Table 4-3 *Oxygen demand and suspended solids in sewage effluents*

Type of return water	5-day, 20°C BOD	Suspended solids, mg/liter
Raw sewage	200	300
Primary settled sewage	130	150
Secondary effluent	20-30	20

Thus it is evident that the return water from domestic use varies in nature in accord with imposed requirements of treatment. A rough estimate of the contribution of quality factors to the water resource pool by domestic return water may be drawn from random statistics. For example, the withdrawals of water for domestic use in 1965 were about 25 billion gallons. At about the same time, the Ohio River Valley Water Sanitation Commission reported [3] data on sewage-treatment facilities (Table 4-4) for the combined states of Illinois, Indiana, Kentucky, New York, Ohio, Pennsylvania, Virginia, and West Virginia.

Table 4-4 *Municipal and
institutional sewage-treatment facilities*

	Percent of sewered communities	Percent of population served
Control currently acceptable	63.5	80.3
Treatment provided, improvements under construction	1.6	6.3
Treatment provided, improvements needed	5.8	3.2
New treatment works under construction	7.0	4.3
No treatment	22.1	5.9

SOURCE: Ohio River Valley Water Sanitation Commission [3].

These data probably mean that about 94 percent of the sewage from the urban population is now being given primary or secondary treatment, whereas 6 percent is discharged raw.

Putting the load in terms of BOD to the resource pool, in 1960 the Senate Select Committee [4] estimated the combined organic load from domestic and industrial use as shown in Table 4-5.

Table 4-5 *Organic loads from
municipal wastes*

Source	1954	1980	2000
Population served, million	100	192.6	279.4
Industries served, million P.E.*	36	96.8	139.7
Total municipal load, million P.E.*	136	289.4	419.1
Removal of BOD, percent	44	70.0	80.0
Residual BOD to water resource, million P.E.*	75	86.8	83.8

*P.E. means population equivalent, based on 100 gal with BOD of 200 mg/liter
 = 1 person.

From this table it is estimated that the combined BOD returned to the water resource by domestic and industrial use in 1980 will be equivalent to the raw sewage from 86.8 million people.

Assuming that the 70 percent reduction in BOD is attained by 1980 through a great increase in the number of secondary plants, that the value applies equally to domestic and industrial wastes, and that the load distribution between domestic and industrial load is as indicated in the table for 1980, we may compute the division of the 86.6×10^6 to be approximately:

(66%) 50×10^6 ascribable to domestic return water

(34%) 29×10^6 ascribable to industrial wastes

By a straight-line interpolation of the values in the foregoing table, 1965 values would appear to be:

BOD removal: 55%

Residual BOD P.E. $= 80 \times 10^6$ people

55×10^6 (69%) ascribable to domestic return water

25×10^6 (31%) ascribable to industrial return water

Similar calculations may be made from data presented on page 8 of a Committee Report [5] of the Eighty-eighth Congress. Here it is shown that in 1960, 83.6 percent of the 110 million population served by sewers also had treatment at some level or another; 18×10^6 people discharged raw sewage; while the population equivalent of the total discharge (return flow) from domestic and industrial users was 75×10^6.

Other interesting data abstracted from Appendix I of the Committee Report are shown in Table 4-6.

Table 4-6 *Population of community versus waste treatment*

Population size group	Discharging untreated wastes		Discharging inadequately treated wastes	
	Number	Population, millions	Number	Population, millions
Under 500	233	0.07	184	0.06
500–1,000	284	0.21	279	0.2
1,000–5,000	742	1.6	712	1.6
5,000–10,000	151	1.03	156	1.06
10,000–25,000	85	1.3	122	2.8
25,000–50,000	25	0.9	31	1.2
50,000–100,000	20	1.9	21	1.34
Over 100,000	10	6.04	30	8.8

The approximate values presented in the table are indicative of the scale of the water quality affecting potential of water returned to the resource pool for domestic use.

Considering again the Committee Report [5], plus the fact that some 26×10^6 people used septic tanks in 1960, we might estimate:

110×10^6 @ 83.6% = 92×10^6 with primary or secondary treatment (68%)

18×10^6 without treatment (13%)

26×10^6 with septic tanks (30×10^6 in 1964) (19%)

136×10^6 urban dwellers

44×10^6 rural (probably not more than 30×10^6)

180×10^6 total population

A rough estimate of the magnitude of the water quality management problem resulting from domestic use can be made from the foregoing and other available data. Assuming, for example, that the 1980 pollutional load from domestic use is indeed equivalent to the raw sewage of 58×10^6 people, and that the 5-day BOD of sewage is 200 mg/liter, the oxygen demand of domestic return water would be some 96×10^5 lb of oxygen. This is equal to the oxygen resource of about 144×10^9 gallons of freshwater having a dissolved oxygen content of 8 mg/liter. If 5 mg/liter of oxygen were reserved for aquatic life, some 384×10^9 gal might be required to supply the oxygen deficit. This is more than one-half the 630×10^9 gpd estimated [6] to be an economically feasible minimum flow which might be available to the United States by use of storage.

There are, of course, discrepancies of scale in comparing the 5-day oxygen demand to the oxygen resource of a daily flow of water. Nevertheless, such a rough comparison is sufficient to support the conclusion that domestic use of a very small fraction of the freshwater resource, even with great improvement in the efficiency of conventional waste treatment, has very serious implications in terms of water quality, and that in the management of water quality, engineers must look seriously at the interchange between domestic use and the resource pool.

REFERENCES

1 H. E. Babbitt and E. R. Baumann: *Sewerage and Sewage Treatment*, John Wiley & Sons, Inc., New York, 1958.

2 *Standard Methods for the Examination of Water and Waste Water*, 12th ed., American Public Health Association, 1965.

3 *Sixteenth Annual Report, ORSANCO*, Ohio River Valley Water Sanitation Commission, 1964.

4 *Water Resources Activities in the United States: Pollution Abatement*, Senate Select Committee Print No. 9, 86th Cong., 2d Sess., 1960.

5 *A Study of Pollution—Water*, Committee Print, 88th Cong., 1st Sess. (Staff Report to the Committee on Public Works, 1963.)

6 W. L. Picton: "Water Use in the United States, 1900–1980," Business and Defense Services Administration, U.S. Department of Commerce, 1960.

CHAPTER 5

Quality changes by industrial use

COMPARISON OF INDUSTRIAL AND DOMESTIC WASTE WATERS

Industrial use of water has quality effects similar to those of domestic use in that it increases the concentration of dissolved and suspended matter in water. Here, however, the similarity ends because of the vast variety of industrial wastes. Essentially only food and fiber processing produces residues that are characteristic of nature's cycle of growth and decay, and hence subject to the same biodegradation as domestic sewage. It has been estimated that two-thirds of the pollution of United States waters derives from industry, whereas Table 4-5 indicates that the population equivalent of industry's organic load totals only about one-half that of domestic use. Many industries discharge process waters carrying compounds that are never found in natural waters, that do not exist in nature, and that occur in a vast spectrum of ever-changing variety. Metal ions (mostly toxic), exotic organic and inorganic chemicals, and many refractory compounds are among the most significant. Waste streams with high temperature, turbidity, color, acidity, or alkalinity likewise are common in industrial use of water.

Another significant difference between industrial and domestic use of water is the percentage of water consumed. Some 66 percent (10 to 95 percent) of industrial water use is for cooling purposes. About half of this is lost to the atmosphere, and the remaining half returns to the resource pool with its salt concentration doubled. This is an average value, however, as the consumption of industrial water by evaporation varies from 0 to 85 percent of cooling water in different industries. Heavy users of cooling water, such as the petrochemical industry, often locate along the seacoast where they make use of saline waters for cooling purposes and discharge the return water to the saline resource with little, if any, quality effect on the freshwater resource. This, as noted in Chapter 7, by no means indicates a freedom from water quality management problems although the effect may be less severe.

In inland areas heavy water-using industries locate along the major rivers in order to obtain cooling water. The 1966 Yearbook of the Ohio River Valley Water Sanitation Commission (ORSANCO) [1] notes that 87 major industries which are self-supplied from the Ohio River in the six partici-pating states* withdraw fifty times as much water for the nonconsumptive use of cooling as do 130 communities (total population more than 2×10^6) for public water supply. Such cooling water represents about 98.7 percent of the total industrial withdrawals. The other 1.3 percent, used as process and boiler makeup water, however, represents some 185 mgd—in compari-son with 276.4 mgd withdrawn for domestic use by the 130 communities.

SOME FACTORS
OF SCALE

A summary of the past and anticipated increase in industrial use, and its comparison with similar increases in other beneficial uses, may be obtained from Table 5-1 prepared by the U.S. Department of Commerce [2]. The

*Illinois, Indiana, Kentucky, Ohio, Pennsylvania, and West Virginia.

Table 5-1 *United States: estimated water use, 1900-1975 (Billions of gallons, daily average)*

Year	Irriga-tion	Public water supplies	Domes-tic	Self-supplied Industrial and miscellaneous	Steam-elec-tric power	Total
1900	20.2	3.0	2.0	10.0	5.0	40.2
1910	39.0	4.7	2.2	14.0	6.5	66.4
1920	55.9	6.0	2.4	18.0	10.0	92.3
1930	60.2	8.0	2.9	21.0	18.4	110.5
1940	71.0	10.1	3.1	29.0	22.2	135.4
1944	80.6	12.0	3.2	56.0	35.9	187.7
1945	83.1	12.0	3.2	48.0	28.8	175.1
1946	86.4	12.0	3.5	39.0	26.9	167.8
1950	100.0	14.1	4.6	46.0	38.4	203.1
1955	119.8	17.0	5.4	60.0	59.8	262.0
1960	135.0	22.0	6.0	71.9	77.6	312.5
1965	148.1	25.0	6.5	87.7	92.2	359.5
1970	159.0	27.8	6.9	103.0	107.8	404.5
1975	169.7	29.8	7.2	115.4	131.0	453.1

SOURCE: U.S. Department of Commerce [2].

values shown cannot be directly related to the magnitude of the freshwater resource because of recycling in some manufacturing processes, saline water inputs, high consumptive use of agriculture, and low consumptive use of public water supplies. Nevertheless, it underscores the rapid rise of industry and power generation as users of water and makes evident the reasons for the change in quality control objectives outlined in Chapter 3.

From Table 5-1 the percentage increase in the decade ending in 1975 may be estimated as shown in Table 5-2.

Table 5-2 *Percent increase in water use by decades*

Year	Irrigation	Public water supply	Indus-trial	Steam-electric power	Total
1955					
1965	23	47	46	54	37
1975	15	20	32	42	27
1975 in terms of 1965	115	120	132	142	127

Similar estimates of the 1980 demand in terms of 1960 for California are as follows:

Total per capita increase	170%
Industrial use increase	250%
Irrigation use	18%

These tables show clearly the growing importance of industry as a user of water. Consumptive use by steam power is the fastest growing. Its effect is to reduce the resource pool and hence to increase the effect of wastes discharged by other users.

Industry's waste return to the pool may be roughly estimated as:

1. One-third of withdrawal returned with the salt content approximately doubled.

2. One-third of withdrawal contaminated with a spectrum of organic and inorganic solids originating in process waters.

3. One-third of withdrawal consumed by incorporation in product or loss to the atmosphere.

The tables show also that the load on the resource pool in terms of pollutional intensity through consumptive depletion and return of used water is growing some 30 percent per decade.

Because the term "industry" refers to a very broad spectrum of economic activity which involves users whose demand for water ranges from near zero to perhaps eight hundred times the weight of product produced, and which mistreats this water in myriad ways, industrial use cannot be estimated by any unit such as the "per capita" applicable to domestic usage. Thus its effect on the quality of the water resource can be judged only in relation to some specific sector of the resource and according to the size and nature of the individual industries involved. Nevertheless, some general idea of the scale of quality change by industrial use can be gained from a consideration of the amounts of water required by various industrial processes and the kinds of quality factors added in the process.

WATER REQUIREMENTS OF INDUSTRY

One distinguishing feature of industrial use of water, in contrast with domestic use, is the extreme variability in water usage from one industry to another. A government survey reported [3] in 1966, summarized in Table 5-3, gives an idea of both this variability and the overall magnitude of industry's use of the overall water resource.

Table 5-3 *Annual water use by selected types of industry*

Type of industry	Annual use, gal	Approximate percent of total industrial use
Primary-metal manufacture	4.5×10^{12}	32
Chemicals and allied products	3.2×10^{12}	23
Paper industry	2×10^{12}	14
Petroleum and coal products	1.3×10^{2}	9
Food industry	750×10^{9}	5
Stone and vitreous products	248×10^{9}	2
Transportation equipment	247×10^{9}	2
Rubber and plastics	163×10^{9}	1
Nonelectrical machinery	103×10^{9}	
Fabricated metals	57×10^{9}	
Instruments and related products	30×10^{9}	
Leather	16×10^{9}	
Miscellaneous	13×10^{9}	
Tobacco	3×10^{9}	
Furniture	3×10^{9}	

SOURCE: American Water Works Association [3]. Reproduced by permission.

Incomplete data compiled from a number of sources [4, 5, 6] and presented in Table 5-4 relate water usage to unit of product. They show that in many cases the ratio of water withdrawn to product produced is extremely large.

Table 5-4 *Water requirements per unit of product, selected U.S. industries*

Industry and product	Unit of product, tons (except as noted)	Water required per unit, U.S. gallons
Chemical industries		
Acetic acid	Ton of HAc	120,000–290,000
Alcohol	Gallon	52–138
Ammonia	Ton of NH_3	37,500
Ammonium sulfate	. . .	240,000
Calcium carbide	. . .	36,500
Calcium metaphosphate	Ton of $Ca(Po_3)_2$	2,800
Carbon dioxide	. . .	24,500
Caustic soda	Ton of NaOH (11%)	22,000–25,500
Cellulose nitrate	. . .	12,000
Charcoal and wood chemicals	Ton of $CaAc_2$	79,000
Corn refining	Ton of starch	333
Gasoline	Gallon	7–34
Gunpowder	. . .	200,000
Hydrochloric acid	Ton of 20 Bé° HCl	3,500
Hydrogen	Ton of H_2	800,000
Lactose	. . .	235,000
Oxygen	100 ft³	65
Soap	. . .	300–600
Soda ash	. . .	18,000–22,000
Sulfuric acid	Ton of 100% H_2SO_4	800–6,000
Sulfur	. . .	3,000
Food and beverage industries		
Beet sugar	. . .	20,000–25,000
Bread	. . .	600–1,200
Beans, green	. . .	20,000
Peaches and pears	. . .	5,300
Other fruits and vegetables	. . .	2,000–10,000
Gelatin	. . .	15,000–24,000
Meat packing	Ton live weight	5,000
Milk products	. . .	4,000–5,000
Oils, edible	Gallon	88
Sugar	. . .	1,200–2,600
Beer	Gallon	15
Whisky	Gallon	80

Table 5-4 *Water requirements per
unit of product, selected U.S.
industries (Continued)*

Industry and product	Unit of product, tons (except as noted)	Water required per unit, U.S. gallons
Pulp and paper		
Kraft pulp	Ton dry pulp	110,000
Sulfate pulp	. . .	82,000
Sulfite pulp	. . .	82,000–230,000
Soda pulp	. . .	101,000
Paper	. . .	47,000
Paperboard	. . .	17,500–103,000
Strawboard	. . .	31,500
Textile industries		
Cotton	. . .	20,000–76,000
Cotton bleaching	. . .	72,000–96,000
Cotton dyeing	. . .	9,500–19,000
Linen	. . .	200,000
Rayon	Ton of yarn	105,000–240,000
Wool scouring	. . .	40,000–240,000
Metal and metal products		
Rolled steel	. . .	96,000
Finished steel	. . .	79,000
Fabricated steel	. . .	52,000
Steel sheets	. . .	16,000
(Average all products)	. . .	20,000–35,000
Aluminum	. . .	360,000
Miscellaneous		
Electric power	kwhr	85–185
Coal washing	. . .	1,800–4,300
Leather tanning	Ton, raw hide	19,500
Synthetic rubber	. . .	24,000–800,000

In evaluating Table 5-4 in terms of water quality management, it should be recalled that the given values represent data derived from industrial practices common before quality of receiving waters placed an important constraint upon industry. Like many data which get into the literature and persist for decades, their validity decreases as the inefficient practice on which they were based gives way to competition or is altered in response to water quality control regulations. It is quite probable that the pulp and paper industry, for example, in proposing a kraft pulp installation in a water-short area in 1967, would design on the basis of something like

20,000 gal/ton instead of the 110,000-gal value given in the table. The fact is that many industrial plants today are neither as good as may be expected in the future nor as bad as those of the past. On the other hand, in some cases the water use per unit of product continues to increase as a result of process changes in spite of greater attention to waste-water quality requirements. The table, therefore, gives some scale to individual industrial demands on the water resource and identifies those types which require large or small amounts of water, even though values shown are not suitable for design purposes.

The use of water by industry varies markedly with the water quality objectives it must meet in a local situation. Granted that industry tends to locate where water supply is a favorable factor, there are times when other raw materials are the controlling consideration both in location of plant and in design of process. One of the best known examples of this latter situation is the Kaiser Steel Plant at Fontana, California. Here the water demand per ton of steel is some 1,400 gal in contrast with the widely quoted figure of 64,000 gal for the industry. However, the 1,400 gal is consumed by loss to the atmosphere and there is no return flow from the plant to the water resource such as might come from the 64,000-gal usage. In this case there is no water resource to which used water might be returned. The plant was located where ore and minerals were favorable, and was designed around the reality that any receiving water would be 100 percent industrial waste, quickly reaching the groundwater by infiltration and percolation.

In contrast, the heavy industry of the Ohio River Valley located where both raw materials and water resources were favorable. In this case it is estimated [1] that even under drought conditions the combined industrial and domestic return flows constitute less than 20 percent of river flow.

NATURE OF
INDUSTRIAL WASTES

The general types of industrial wastes, together with the principal industries which produce each type, are summarized in Tables 5-5 and 5-6.

The biological wastes from industry are particularly significant in comparison with domestic wastes because of the exceptionally high BOD of many such discharges. It is generally estimated that the average BOD of organic industrial discharges, when process, washdown, and other return streams from the industrial plant are combined, is of the order of 3,500 mg/liter—more than ten times that of a strong domestic sewage. Averages, however, are often meaningless because of the diversity of industrial use.

Some concept of the oxygen demand of various industrial activities may be obtained from Table 5-7. The values in this table do not, of course, apply directly to the total withdrawals summarized in Table 5-4, which includes

Table 5-5 *Types of industrial wastes and principal industries producing each type*

Chiefly mineral, or partly mineral–partly organic materials	Chiefly organic materials
Brine wastes Mineral washings, e.g., stone sawing, sand and china clay wash Mine drainage (coal pit water) Pickle liquor (Fe, Cu, and Zn) Electroplating Water softening Cooling water Boiler blow off Inorganic chemical wastes Battery manufacture Inorganic pigments Coal washing Photographic wastes	Hydrocarbons: Oil wells Petroleum refining Styrene manufacture Butadiene plants Processing natural rubber Gasoline stations, garages Copolymer rubber plants Miscellaneous organic chemicals: Munition plants Synthetic pharmaceuticals Synthetic fibers Organic chemical manufacture Paints and varnishes Oil and grease processing Phenolic wastes: Gas and coke byproducts Tar distillation and creosoting Chemical plants Synthetic resin plants Wood distillation Dye manufacture Biological wastes: *Biological processing* Tanneries and leather trades Pharmaceuticals Alcohol industries Misc. fermentation industries Glue, size, and gelatin plants Wool scouring Textile manufacture Floor cloth manufacture Paper manufacture Laundries *Food processing* Canneries Meat packing, etc. Milk and dairy wastes Corn products plants Beet sugar factories Cane sugar factories Fish processing Food dehydration

SOURCE: Ettinger [7]. Reproduced by permission of the publisher.

Table 5-6   *Some significant chemicals in industrial waste waters*

Chemical	Industry
Acetic acid	Acetate rayon, pickle and beetroot manufacture
Alkalies	Cotton and straw kiering, cotton manufacture, mercerizing, wool scouring, laundries
Ammonia	Gas and coke manufacture, chemical manufacture
Arsenic	Sheep-dipping, fell mongering
Chlorine	Laundries, paper mills, textile bleaching
Chromium	Plating, chrome tanning, aluminum anodizing
Cadmium	Plating
Citric acid	Soft drinks and citrus fruit processing
Copper	Plating, pickling, rayon manufacture
Cyanides	Plating, metal cleaning, case-hardening, gas manufacture
Fats, oils, grease	Wool scouring, laundries, textiles, oil refineries
Fluorides	Gas and coke manufacture, chemical manufacture, fertilizer plants, transistor manufacture, metal refining, ceramic plants, glass etching
Formalin	Manufacture of synthetic resins and penicillin
Hydrocarbons	Petrochemical and rubber factories
Hydrogen peroxide	Textile bleaching, rocket motor testing
Lead	Battery manufacture, lead mining, paint manufacture, gasoline manufacture
Mercaptans	Oil refining, pulp mills
Mineral acids	Chemical manufacture, mines, Fe and Cu pickling, DDT manufacture, brewing, textiles, photo-engraving, battery manufacture
Nickel	Plating
Nitro compounds	Explosives and chemical works
Organic acids	Distilleries and fermentation plants
Phenols	Gas and coke manufacture; synthetic resin manufacture; textiles; tanneries; tar, chemical, and dye manufacture; sheep-dipping
Silver	Plating, photography
Starch	Food, textile, wallpaper manufacture
Sugars	Dairies, foods, sugar refining, preserves, wood process
Sulfides	Textiles, tanneries, gas manufacture, rayon manufacture
Sulfites	Wood process, viscose manufacture, bleaching
Tannic acid	Tanning, sawmills
Tartaric acid	Dyeing; wine, leather, and chemical manufacture
Zinc	Galvanizing, plating, viscose manufacture, rubber process

SOURCE: Klein [8]. Reproduced by permission of the publisher.

Table 5-7 *BOD of wastes from selected industries*

Source of waste	5-day, 20°C BOD of waste, mg/liter
Beet sugar refining	450–2,000
Brewery	500–1,200
Beer slop	11,500
Cannery	300–4,000
Grain distilling	15,000–20,000
Molasses distilling	20,000–30,000
Laundry	300–1,000
Milk processing	300–2,000
Meat packing	600–2,000
Pulp and paper	
Sulfite	20
Sulfite-cooker	16,000–25,000
Tannery	500–5,000
Textiles	
Cotton processing	50–1,750
Wool scouring	200–10,000

consumptive and cooling water use as well as process water. Neither do they apply to individual process streams within the plant which may require special treatment before discharge with the plant effluent. Such waste streams may be several times as strong, in terms of oxygen demand, as the values shown in the table. They do, however, call attention to the disparity between domestic and organic industrial wastes, as well as the need to express the strength of such return flows in terms of population equivalent (P.E.), as in Table 4-5.

EFFECT OF INDUSTRIAL WASTES

The consequence of industrial use of water, in quantities such as specified in Tables 5-1 and 5-4, is the wide variety of quality-degrading factors summarized in Tables 5-5, 5-6, and 5-9. Some of these factors are in themselves toxic to aquatic life and higher animals. Others can create aesthetically objectionable conditions. Still others might interfere with various uses by reason of minerals or biochemically unstable organic matter.

Both toxicity and oxygen depletion are of concern where protection of aquatic life is an objective of a resource-oriented policy of water quality management. The public is particularly sensitive to fish kills, and wastes in the return waters from industry have been reported [9] to be most responsible.

The harmful effects of organic and other materials in either domestic or industrial return waters are summarized in Table 5-8 below.

Table 5-9 [10], on pages 52-57, presents a summary of the origin and characteristics of the major wastes in a number of industries, as well as of the principal methods utilized in quality management. The table may serve to suggest points in the industrial activity itself where engineered subsystems may serve the goals of water quality management. The merits and limitations of the major treatment and disposal methods listed in the table are appraised in a later chapter.

Table 5-8   *Harmful effects of domestic and industrial wastes*

Type of material	Effect
Biodegradable organic matter	Deoxygenate water; kill fish, objectionable odors.
Suspended matter	Deposit on riverbed; if organic, may putrify and float masses to surface by gas; blanket bottom and interfere with fish spawning or disrupt food chain.
Corrosive substances (e.g., cyanides, phenols, metal ions)	May kill fish and other aquatic life; destroy bacteria and so interrupt self-purification of streams.
Pathogenic micro-organisms	Sewage may carry pathogens; tannery wastes, anthrax.
Substances causing turbidity, temperature, color, odor, etc.	Temperature rise may injure fish; color, odor, turbidity may render water aesthetically unacceptable for public use.
Substances or factors which upset biological balance	May cause excessive growth of fungi or aquatic plants which choke stream, cause odors, etc.
Mineral constituents	Increase hardness, limit use in industry without special treatment, increase salt content to level deleterious to fish or vegetation, lead to eutrophication of water.

SOURCE: Klein [8]. Reproduced by permission of the publisher.

STATUS OF QUALITY MANAGEMENT OF INDUSTRIAL WASTES

Currently, it is essentially impossible to amass reliable figures on either the potential or the present contribution of inorganic or exotic inorganic compounds to the water resource pool via industrial return water. Production figures of varying degree of reliability can be obtained, but inplant practices are variable within any single industry as well as from industry to industry. Consequently, the whole matter of waste discharges from industry is only partially investigated, understood, or policed. Water pollution control agencies have, of course, assembled a vast amount of data in the pollution control activity of the past 15 years. The 1964 ORSANCO report [11] showed that about 82 percent of industries in the eight member states had currently acceptable control facilities, and another 17 percent had either some type of facility or were in various stages of planning or construction. In all, 89.7 percent were complying with ORSANCO minimum requirements.

Several factors combine to obscure the national picture. First, as has been previously pointed out, the emphasis on water quality management has only recently shifted from pollution control to management of the quality of the water resource. To this end, Federal legislation established a 1967 deadline for standards of quality to be set up by the states, or by the Federal government should the states fail to act. At the same time, industrial use of water is growing apace as is the variety of goods produced. There is a general impression that what is being done is not enough and that technology and economic concepts lag behind the needs of quality management. The one certain fact is that a very major part of the water quality management activity pursuant to the quality goals of the future will concern the management of water in industrial use.

REFERENCES

1 *Eighteenth Yearbook 1966,* Report of Commissioners, Ohio River Valley Water Sanitation Commission, Cincinnati, Ohio.
2 "Water Use in the United States, 1960-1975," U.S. Department of Commerce, BSB-136, January, 1956.
3 "Water Control News, Commerce Clearing House," *Willing Water,* American Water Works Association, August, 1966.
4 R. B. Canlan: "Water Re-use and Conservation," *Betz Indicator,* 29(12), December, 1960.
5 George E. Symons: "The Disposal of Industrial Wastes," *Water Works & Sewerage,* 92(12), December, 1945.
6 "Water for Industrial Use," United Nations, 1958.
7 M. B. Ettinger: "Analytical Procedures for Industrial Wastes," *Water Sewage Works,* 97(7), July, 1950.
8 Louis Klein: *River Pollution. 2: Causes and Effects,* Butterworth & Co. (Publishers), Ltd., London, 1962.
9 "Industrial Wastes Blamed for Most Fish Kills" (Summary of U.S. Public Health Service Report), *Water Sewage Works,* 108(10), October, 1961.
10 N. L. Nemerow: *Theory and Practice of Industrial Waste Treatment,* Addison-Wesley Publishing Company, Inc., Reading, Mass., 1963.
11 *Sixteenth Annual Report, ORSANCO,* Ohio River Valley Water Sanitation Commission, Cincinnati, Ohio, 1964.

Table 5-9 *Summary of industrial waste: its origin, character, and treatment*

Industries producing wastes	Origin of major wastes	Major characteristics	Major treatment and disposal methods
		Food and drugs	
Canned goods	Trimming, culling, juicing, and blanching of fruits and vegetables	High in suspended solids, colloidal and dissolved organic matter	Screening, lagooning, soil absorption or spray irrigation
Dairy products	Dilutions of whole milk, separated milk, buttermilk, and whey	High in dissolved organic matter, mainly protein, fat, and lactose	Biological treatment, aeration, trickling filtration, activated sludge
Brewed and distilled beverages	Steeping and pressing of grain, residue from distillation of alcohol, condensate from stillage evaporation	High in dissolved organic solids, containing nitrogen and fermented starches or their products	Recovery, concentration by centrifugation and evaporation, trickling filtration; use in feeds
Meat and poultry products	Stockyards, slaughtering of animals, rendering of bones and fats, residues in condensates, grease and wash water, picking of chickens	High in dissolved and suspended organic matter, blood, other proteins, and fats	Screening, settling and/or flotation, trickling filtration
Beet sugar	Transfer, screening and juicing waters, draining from lime sludge, condensates after evaporator, juice, extracted sugar	High in dissolved and suspended organic matter, containing sugar and protein	Reuse of wastes, coagulation, and lagooning

Pharmaceutical products	Mycelium, spent filtrate, and wash waters	High in suspended and dissolved organic matter, including vitamins	Evaporation and drying, feeds
Yeast	Residue from yeast filtration	High in solids (mainly organic) and BOD	Anaerobic digestion, trickling filtration
Pickles	Lime water; brine, alum and turmeric, syrup, seeds and pieces of cucumber	Variable pH, high suspended solids, color, and organic matter	Good housekeeping, screening, equalization
Coffee	Pulping and fermenting of coffee bean	High BOD and suspended solids	Screening, settling, and trickling filtration
Fish	Rejects from centrifuge, pressed fish, evaporator and other wash water wastes	Very high BOD, total organic solids, and odor	Evaporation of total waste, barge remainder to sea
Rice	Soaking, cooking, and washing of rice	High in BOD, total and suspended solids (mainly starch)	Lime coagulation, digestion
Soft drinks	Bottle washing, floor and equipment cleaning, syrup-storage-tank drains	High pH, suspended solids and BOD	Screening, plus discharge to municipal sewer
Apparel			
Textiles	Cooking of fibers, desizing of fabric	Highly alkaline, colored, high BOD and temperature, high suspended solids	Neutralization, chemical precipitation, biological treatment, aeration and/or trickling filtration

Table 5-9 *Summary of industrial waste: its origin, character, and treatment (Continued)*

Industries producing wastes	Origin of major wastes	Major characteristics	Major treatment and disposal methods
Apparel (Continued)			
Leather goods	Unhairing, soaking, deliming and bating of hides	High total solids, hardness, salt, sulfides, chromium, pH precipitated lime and BOD	Equalization, sedimentation, and biological treatment
Laundry trades	Washing of fabrics	High turbidity, alkalinity, and organic solids	Screening, chemical precipitation, flotation, and adsorption
Chemicals			
Acids	Dilute wash waters; many varied dilute acids	Low pH, low organic content	Upflow or straight neutralization, burning when some organic matter is present
Detergents	Washing and purifying soaps and detergents	High in BOD and saponified soaps	Flotation and skimming, precipitation with $CaCl_2$
Cornstarch	Evaporator condensate, syrup from final washes, wastes from "bottling up" process	High BOD and dissolved organic matter; mainly starch and related material	Equalization, biological filtration
Explosives	Washing TNT and guncotton for purification, washing and pickling of cartridges	TNT, colored, acid, odorous, and contains organic acids and alcohol from powder and cotton, metal, acid, oils, and soaps	Flotation, chemical precipitation, biological treatment, aeration, chlorination of TNT, neutralization

Insecticides	Washing and purification products such as 2,4-D and DDT	High organic matter, benzene ring structure, toxic to bacteria and fish, acid	Dilution, storage, activated carbon adsorption, alkaline chlorination
Phosphate and phosphorus	Washing, screening, floating rock, condenser bleed-off from phosphate reduction plant	Clays, slimes and tall oils, low pH, high suspended solids, phosphorus, silica, and fluoride	Lagooning, mechanical clarification, coagulation and settling of refined waste
Formaldehyde	Residues from manufacturing synthetic resins, and from dyeing synthetic fibers	Normally has high BOD and HCHO, toxic to bacteria in high concentrations	Trickling filtration, adsorption on activated charcoal

Materials

Pulp and paper	Cooking, refining, washing of fibers, screening of paper pulp	High or low pH; colored; high suspended, colloidal, and dissolved solids; inorganic fillers	Settling, lagooning, biological treatment, aeration, recovery of byproducts
Photographic products	Spent solutions of developer and fixer	Alkaline, contains various organic and inorganic reducing agents	Recovery of silver, plus discharge of wastes into municipal sewer
Steel	Coking of coal, washing of blast-furnace flue gases, and pickling of steel	Low pH, acids, cyanogen, phenol, ore, coke, limestone, alkali, oils, mill scale, and fine suspended solids	Neutralization, recovery and reuse, chemical coagulation
Metal-plated products	Stripping of oxides, cleaning and plating of metals	Acid, metals, toxic, low volume, mainly mineral matter	Alkaline chlorination of cyanide, reduction and precipitation of chromium, and lime precipitation of other metals

Table 5-9 *Summary of industrial waste: its origin, character, and treatment (Continued)*

Materials (Continued)

Industries producing wastes	Origin of major wastes	Major characteristics	Major treatment and disposal methods
Iron-foundry products	Wasting of used sand by hydraulic discharge	High suspended solids, mainly sand; some clay and coal	Selective screening, drying of reclaimed sand
Oil	Drilling muds, salt, oil, and some natural gas, acid sludges and miscellaneous oils from refining	High dissolved salts from field, high BOD, odor, phenol, and sulfur compounds from refinery	Diversion, recovery, injection of salts; acidification and burning of alkaline sludges
Rubber	Washing of latex, coagulated rubber, exuded impurities from crude rubber	High BOD and odor, high suspended solids, variable pH, high chlorides	Aeration, chlorination, sulfonation, biological treatment
Glass	Polishing and cleaning of glass	Red color, alkaline non-settleable suspended solids	Calcium chloride precipitation
Naval stores	Washing of stumps, drop solution, solvent recovery, and oil recovery water	Acid, high BOD	Byproduct recovery, equalization, recirculation and reuse, trickling filtration

Energy

Steam power	Cooling water, boiler blowdown, coal drainage	Hot, high volume, high inorganic and dissolved solids	Cooling by aeration, storage of ashes, neutralization of excess acid wastes
Coal processing	Cleaning and classification of coal, leaching of sulfur strata with water	High suspended solids, mainly coal; low pH, high H_2SO_4, and $FeSO_4$	Settling, froth flotation, drainage control, and scaling of mines
Nuclear power and radioactive materials	Processing ores, laundering of contaminated clothes, research-lab wastes, processing of fuel, powerplant cooling waters	Radioactive elements; can be very acid and "hot"	Concentration and containing, or dilution and dispersion

SOURCE: Nemerow [10]. Reproduced by permission of the publisher.

CHAPTER 6

Quality changes
through agricultural use

INTRODUCTION

It has been said that every civilization that has depended upon irrigated agriculture for its existence has failed. In the United States we like to tell ourselves that our superior knowledge of the soil-water-crop relationship ensures immunity from the fate of Mesopotamia. And perhaps we are right. It does seem unlikely that by our irrigation practices we shall kill the land as did our forebears. Nevertheless, the possibility exists that by killing the water we may achieve the same ultimate end. If such should be the case it may well be agricultural and industrial return waters which overwhelm the water resource. One of the goals of water quality management, therefore, must be to preclude such an eventuality by an awareness of the nature and scale of quality-related factors introduced into the freshwater resource pool by return water from agricultural use and by appropriate action.

Historically [1], federally financed irrigation water was made possible by the Homestead Act of 1902, designed to put self-sustaining families on the semiarid lands of the 17 Western states. One of the precursors of aridity, however, is the absence of adequate rainfall to leach soluble salts from the soil or to carry away those brought up by capillarity or rejected by plants in the root zone. Irrigation water must therefore do this task; hence mineralization becomes the chief quality attribute of irrigation return waters. The problem is compounded by several phenomena. First, water for agricultural use must be impounded and stored under a blazing sun that reclaims 6 to 10 ft of water per year from reservoirs in the Southwest. Thus, its salinity is increased and its leaching capacity reduced, requiring increasing application of irrigation water as time goes on. The result is an increase in both the quantity of water required by irrigated agriculture and in the dissolved solids in its return waters. Furthermore, return waters are further concentrated by evaporative losses. Some percolate to the groundwater to be withdrawn again for agricultural use and a new cycle of salt concentration. It is thus that the long-term prospect of irrigation may be a failure of water quality rather than soil quality.

It is not in arid regions alone, however, that agriculture makes use of irrigation. Supplementary irrigation is becoming a widespread practice in the humid areas of the United States where water applied at the critical stage of plant growth is far more important than the fact of plentiful rainfall on an annual or seasonal basis. Hence the volume of water withdrawn for agricultural use seems destined to increase.

**MAGNITUDE
OF IRRIGATION PRACTICE**

Table 6-1 [2] presents data regarding the magnitude and distribution of irrigation in terms of acreage and acreage increase for the 2-year period 1954 to 1956. Several notable facts from this table are evident:

1. A 17 percent increase in irrigated acreage occurred in the 2-year period. Almost 5.5×10^6 acres were added, with about 4.6×10^6 of this total occurring in the 17 Western states.
2. About 41 percent of the land under irrigation is in California and Texas.
3. Six states, five of which are in the humid areas of the United States, increased irrigation acreage over 100 percent, signifying the emergence of supplemental irrigation as a phenomenon in American agriculture.

It is estimated that ultimately some 51×10^6 acres will be under irrigation in the 17 Western states; and eventually, supplemental irrigation may be applied to much of the 125×10^6 irrigable acres in the humid Eastern states.

**QUANTITY
OF IRRIGATION WATER**

Estimates of the total withdrawals of irrigation water vary from one reference to another but are generally consistent. In 1949 it was estimated [3] that 95.44×10^6 acre-ft of irrigation water was used during that year (75 percent from surface supplies and 25 percent from groundwater).

A 1962 estimate by Wolman [4] placed irrigation use at 159×10^6 acre-ft/year at that time. Public Health Service projections [5] in 1960 anticipated a use of 165.7×10^9 gpd (185×10^6 acre-ft/year) in 1980 — a value somewhat smaller than the 169.7×10^9 gpd (190×10^6 acre-ft/year) estimated by UNESCO in 1958 [6].

For any given area the magnitude of irrigation needs can be computed from the data available locally on the rates of application of water. The 1949 Census of Agriculture [3] estimated that the use of water on land in the 17 Western states averaged 3.4 acre-ft/acre/year, with variations from 2.2 in the Rio Grande Valley to 6.1 in Idaho. The rates are governed not only by the nature of the crop irrigated, but also by cost of water, soil types, local climatic conditions, and water quality. For humid climates the rates vary from 5.0 to 1.25 acre-ft/acre/year.

Table 6-1   *U.S. irrigated acreage*

State	Acreage 1956	Acreage 1954	Increase, percent	Sprinkler*
U.S. total	36,002,627	30,711,453	17	3,064,463
17 Western states	32,661,501	28,092,947	16	2,010,062
Other states	3,341,126	2,618,506	28	1,054,401
1. California	7,750,000	7,048,792	10	400,000
2. Texas	6,962,234	5,439,000	28	575,015
3. Idaho	2,405,089	2,324,571	3	130,000
4. Colorado	2,382,000	2,263,000	5	33,110
5. Nebraska	2,012,320	1,393,733	44	201,230
6. Montana	1,890,000	1,860,000	. . .	58,000
7. Utah	1,612,108	1,072,682	50	3,325
8. Oregon	1,575,000	1,490,397	6	157,500
9. Wyoming	1,300,000	1,262,632	3	8,000
10. Arizona	1,150,000	1,250,000	−9	1,000
11. Washington	947,000	778,135	22	228,000
12. Arkansas	892,936	857,390	4	54,756
13. Florida	821,282	428,282	92	180,000
14. New Mexico	800,000	649,615	23	3,000
15. Kansas	722,575	420,000	72	100,000
16. Louisiana	711,000	707,818	1	39,650
17. Nevada	700,000	567,498	23	12,000
18. Oklahoma	285,175	107,981	164	100,882
19. Mississippi	157,000	132,490	19	54,000
20. South Dakota	120,000	90,371	33	20,000
Increase of over 100 per cent in:				
Iowa	20,000	2,396	818	20,000
Maine	6,900	1,097	529	6,850
Georgia	80,000	27,701	189	79,200
Oklahoma			164	
Virginia	45,500	21,805	116	45,050
Delaware	11,000	5,553	100	11,000

*31.5 percent of the irrigation in "other states" uses sprinklers, compared to only 6.1 percent in "17 Western states."

SOURCE: Eldridge [2]. Taken from the 1957 Directory and Buyers Guide.

COMPARISON WITH
OTHER MAJOR USES

In Chapter 5 data were presented which show that although irrigation usage is greater than either industrial or domestic, its comparative rate of growth is diminishing. The same pattern is revealed in Table 6-2 based on data [5] released by the U.S. Public Health Service.

Table 6-2 *Pattern of water use in the United States (1960)*

Year	Total water usage, billion gpd	Distribution of water usage, percent		
		Agriculture	Industry	Municipal
1900	40	55.0	37.5	7.5
1960	324	43.6	50.0	6.5
1980	547	27.6	65.6	6.8

A decline in the predominance of agricultural water is evident in the table. Nevertheless, agricultural return waters, and the consumptive loss of water from the resource pool which agriculture occasions, will continue to have a profound effect on the overall quality of the pool.

RETURN IRRIGATION WATER

The disposition of water diverted to agricultural use from surface and from ground sources is illustrated in the flow diagrams in Figure 6-1. There is considerable variability in the quantity of return flow (20 to 60 percent), but the average in the western United States is considered to be about 33 percent of the water diverted. The variation is governed by the same factors that determine rate of application.

In the simple logistics of handling irrigation water a number of losses occur which affect the quality of the water and hence of irrigation return waters. Canal losses vary from 15 to 40 percent of the water diverted; 5 to 25 percent is wasted from canals and laterals; farm wastes, from 5 to 10 percent; and seepage from 5 to 60 percent. Overflows from canals and laterals, runoff from land, and seepage will return water either to the surface- or ground-water sectors of the resource pool at an increased salt concentration due to evaporation. Transpiration by plants growing along irrigation works depletes the supply in quantity and leaves salts available for pickup by percolating water. Evaporation from reservoirs accounts for about 15 million acre-ft/year. This is about 35 percent of the total reservoir capacity available for irrigation. Its effect is to concentrate the salts in the reservoir. Thus it is evident that the waterworks utilized in irrigation themselves increase the salinity of the resource pool.

QUALITY OF IRRIGATION
RETURN WATER

Figure 6-1 identifies three major sources of return water as overflow, runoff, and seepage.

Figure 6-1   *Disposition of water in agricultural use. (Eldridge [2].)*

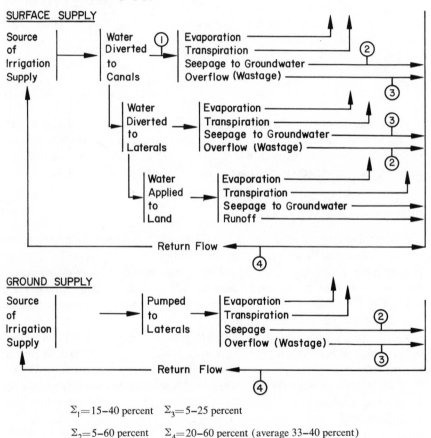

$\Sigma_1 = 15$–40 percent $\Sigma_3 = 5$–25 percent

$\Sigma_2 = 5$–60 percent $\Sigma_4 = 20$–60 percent (average 33–40 percent)

In general the quantity of return water has a significant influence on its quality; the greater the quantity, the less the salt concentration, except where insufficient water applied in one period has left salts in the soil which are removed when more water is applied.

The factors which affect the quality of agricultural return waters include:

1. *Evaporation and transpiration.* Salts contained in the irrigation water are concentrated by the removal of water and retained in this concentrated form in the soil water.

2. *Leaching.* Water must be applied in sufficient amounts to remove excess salts from the soil, and varies from 6 to 25 percent of the water applied.

During the period of reclaiming of soils having a high salinity, 1 ft of water per each foot of soil depth will remove about 80 percent of the salt. Three feet may be necessary to remove boron.

Table 6-3 shows the change in quality of irrigation return waters, as regards salts and ions, from seven areas in the Western states. From the observation that about one-third of the irrigation water returns to the resource pool, it might be expected that the salt concentration is multiplied by 3 unless there is a significant quantity of the salts already contained in the soil or subsurface water. In all the examples summarized in the table the overall increase in salinity was greater than three times. In fact, the lowest is five times. This means that either:

1. The water was used more than once, or
2. The return flow was less than 33 percent, or

Table 6-3 *Increase in salts and ions as a result of irrigation*

Location*	Number of times greater than in irrigation water						
	Salinity	Ca	Mg	Na	HCO$_3$	SO$_4$	Cl
1	10.8	5.1	7.5	21.4	2.4	10.5	129.0
2	7.0	...	...	...	4.3	4.7	3.0
3	5.0	4.4	4.2	8.1	1.8	1.9	
4	5.9	3.7	5.5	14.7	1.8	8.6	11.5
5	7.1	4.6	6.9	12.7	2.9	4.6	40.0
6	5.6	4.4	14.0	3.5	1.6	15.0	Decrease
7	7.8	5.3	6.0	17.2	5.2	18.0	8.0

*The numbers in this column are defined as follows:
1 Rio Grande between Otowi Bridge and Fort Quitman.
2 Yakima River between Easton and Prosser.
3 Average of 1943-44 input and output on an irrigation district.
4 Average of 1940-41 data on Arkansas River between Pueblo and Holly.
5 Average of five years of data on input and output of an irrigated area.
6 Data collected on a short stretch of the Colorado River between Cameo and Grand Junction.
7 Input and output data of an irrigation project in Boise River Basin.
SOURCE: Eldridge [2].

3. A significant portion of salts contained in the soil was dissolved by percolating water, either by simple dissolution or by ion-exchange phenomena involving nitrogen in fertilizers.

Another important change in water quality resulting from irrigation is a shift in the relative concentration of the various ions. The proportion of sodium to calcium and magnesium is always higher in the return water than in the applied water, possibly due to the precipitation of calcium carbonate and to ion exchange. With few exceptions, magnesium remains fairly constant in quantity though changed in concentration.

EFFECT OF RETURN WATER
ON RECEIVING WATER

The observed effects of irrigation return waters as reported by Eldridge [2] are summarized in Table 6-4.

Table 6-4 *Summary of effects*
of agricultural return waters
on receiving waters

Quality factor	Observed effect
Salts and ions	Increased more than five times.
Hardness	Greatly increased; most significant effect.
Total dissolved solids	Greatly increased.
Temperature	Significant increase in Yakima River. Similar effects to be expected in other areas.
Turbidity	High in some return waters. Little studied effect. Small effect in Yakima River due to dilution.
Color	Little studied factor. Increase observed in Yakima River due to large irrigation return.
Nutrients:	Generally somewhat increased.
Nitrates	Conflicting evidence. Evidence of increase in nitrogen but the relative importance of natural and irrigation return flows not investigated.
Phosphorus	May be brought in with plant leaves in return water. Little doubt that amount increased.
Tastes and odors	Evident in return flows as result of mineral salts and organic decomposition products, often from aquatic growth stimulated by nutrient content.
Insecticides and herbicides	Problem exists in surface waters but the relative contribution of irrigation return water and surface runoff not clear. Potential problem exists.
Bacteria	Not a significant factor in agricultural return waters as compared to domestic.

SOURCE: Eldridge [2].

MANAGEMENT OF
AGRICULTURAL
RETURN WATERS

The implications of both primary and supplemental irrigation are of major importance to quality maintenance of the water resource in the semiarid regions of the United States. To begin with, agriculture is a very large user of water, essentially of the same magnitude as industry in 1965. Unlike industry, which tends to locate where water resources are plentiful, irrigated agriculture exists where the resource is minimal. Consequently, dilution is of little significance in upgrading agricultural return waters. On the credit side, mineral pickup is the major quality change from agricultural use, in contrast with the wide range of wastes in industrial return waters and the biochemically unstable organic content of domestic effluents. However, economically feasible technology for demineralizing water is not as advanced as are engineered systems, however inadequate, for managing the quality of industrial and domestic wastes.

A further complication in the management of quality of agricultural effluents is that agriculture is an extractive industry, which, like most extractive industries, is subject to a low multiplier of value of inputs. In fact, irrigated agriculture exists largely as a result of a public policy under which the public and other sectors of the economy underwrite some of the cost of irrigation water. In addition, long-established political attitudes regarding agriculture have combined with economic factors and with a "first-things-first" approach to pollution control to limit concern for quality control of agricultural return waters. Thus the nation approaches the prospect of managing the quality of its water resource with a "clean water" rather than a "pure water" concept that minimizes immediate concern for the mineralizing effect of agricultural use, or of steam-electric power and impoundment for multipurpose uses, for that matter.

This does not mean that the mineral content of return waters from agriculture is to be ignored. The long-term degradation of water quality by this factor is a serious matter to agriculture itself as well as to other beneficial uses. What it means is that we approach the need for water quality management with but little preparation in either policy or technology for upgrading the quality of return waters from agricultural use.

A major example of current concern for irrigation return water is to be found in the currently controversial San Luis Drain in California. Here the vast California Water Plan proposes to convey water south to the head of the San Joaquin Valley to open up new lands to irrigation, salvage already productive land which is running out of groundwater, and supply the needs of other sectors of a burgeoning population. Return waters from such an enterprise will, of course, be largely secondhand water. In the present state of economic technology, desalination is out of the question and quality management must take the form of physical isolation of waste water from the freshwater resource. But even this is not a simple management technique. To return it to San Francisco Bay in the natural bed of the river is to encourage a buildup of salts in groundwater by percolation. On the other hand, a concrete-lined drain is extremely costly. Finally, there is the problem of getting the return stream into the ocean sink. Discharged to the Bay, it disturbs the equilibrium of beneficial use either by encroachment on existing freshwater supplies, or by preempting assimilative capacity now utilized by industry, thus conceivably shifting the entire burden of quality management systems to the industrial and domestic sectors of the economy. Discharged directly to the ocean, at additional cost, other interests are disturbed. Although it may be too salty for the Bay, it is not salty enough for the ocean, and hence becomes a pollutant in the saline environment.

The unfortunate prospect at this point in time is that agricultural use of water will contribute in a major way to the long-term decline in the mineral quality of the water resource in some areas unless better systems for managing the quality of its return waters are forthcoming.

REFERENCES

1 H. Erlich and P. H. McGauhey: *Economic Evaluation of Water: Part I. A Search for Criteria,* Water Resources Contribution No. 13. Sanitary Engineering Research Laboratory, University of California, Berkeley, December, 1957.

2 E. F. Eldridge: "Return Irrigation Water—Characteristics and Effects," U.S. Public Health Service, Region IX, May 1, 1960.

3 U.S. Bureau of the Census, Census of Agriculture, 1950.

4 A. Wolman: "Water Resources: A Report to the Committee on Natural Resources of the National Academy of Science," Publication 1000-B, National Research Council, Washington, D.C., 1962.

5 "Clean Water: A Chart Book of America's Water Needs," U.S. Public Health Service, 1960.

6 "Water for Industrial Use," United Nations Department of Economic and Social Affairs, New York, 1958.

CHAPTER 7

Quality considerations in estuarine waters*

INTRODUCTION

Although the subject of estuarine and marine resources transcends the scope of this book, both the objectives and the techniques of quality management of the freshwater resource must take into account the quality considerations relevant to such waters. This is particularly true of the estuarine waters which, as shown in Chapter 2, are an integral part of the quality interchange system.

For obvious reasons great industrial and commercial cities are located on estuaries and hence millions of people have an intimate concern with the resource values of such bodies of water. Many of these resource values depend upon the quality characteristics of the water; hence, however concerned the citizen may be for "pure" water at the intakes of his freshwater supply, he is never oblivious to the effect of stream discharges on the estuaries which funnel it into the ocean. In fact, the estuary is a more than normally sensitive spot because its aquatic life must always live dangerously. Floods may overwhelm life with freshwater, or shift and obliterate the bottom deposits on which it lives. Diversions may lead to an increase in salinity for lack of flushout; industrial wastes may increase in concentration; or the residence time of domestic effluents increase, with consequent abnormal oxygen depletion.

To the human being, what happens upstream as a result of water quality or quantity management in his behalf may lead to the obliteration of his favorite beach, the disappearance of fish he sought for sport and recreation, increased concentration of pollutants in the estuarine waters, or, as in the case of California's delta area, encroachment of salinity on his most valuable agricultural land. It is therefore pertinent that the engineer concerned with management of the quality of the freshwater resource have some understanding of the constraints which may arise as a result of quality considerations in estuarine waters.

*This chapter co-authored by R. E. Selleck.

WHAT CONSTITUTES
AN ESTUARY

A somewhat poetic definition of an estuary is presented by Webster: "The passage where the tide meets the current of the river." Actually, an embayment is also commonly called an estuary even when there are no significant tributary streams. Often estuaries are classified into three categories: inverse, neutral, and positive. An *inverse* estuary has generally a net landward flux of water, and a *positive* estuary, a net seaward flux. A *neutral* estuary is one which fits into the definition somewhere between the negative and the positive. An estuary is always tidal, hence in the common term "tidal estuary" the "tidal" is redundant.

In the context of water quality management the positive estuary would be most affected by landward control of surface water and water quality, since the seaward flux is a result of freshwater inputs, usually aboveground. The negative estuary might be expected to have the most profound effect on the freshwater resource if its landward flux is the result of saline recharge of underground aquifers as well as of evaporation and transpiration. In any event, it has the least ability of any type of estuary to cope with return flows from beneficial use, although its mere presence is a temptation to beneficial users seeking a cheap transport of wastes to the ocean.

For the purpose of this discussion, both bays and more purely defined passages are considered to be estuaries. Furthermore, in evaluating the quality considerations in estuarine waters it is difficult to exclude entirely the interchange between estuarine and marine waters. Therefore, effects on beach and nearshore waters, beyond which the effects of freshwater pollution are undetectable, are associated with the estuary.

NATURE OF
ESTUARINE RESOURCES

The quality factors which are of importance to estuarine environments and which may be directly concerned in freshwater quality management undertakings are related to aesthetic values, recreation, commercial and sport fishing, industrial water supply, pleasure boating, transportation, and to a lesser extent, oil reserves and underwater mineral deposits. These, at least, represent the principal resource values of estuarine waters which place limitations on the free use of estuaries either as recipients of wastes discharged to influent streams after treatment, or directly into the estuary to avoid mingling of fresh and return water supplies. Associated with some of these resource values are considerations of public health and aquatic ecology which impose further constraints.

A full evaluation of the economic and social worth of estuarine resources in the United States is beyond the scope of this writing. Nevertheless, a few random examples and general relationships, together with more detailed discussion of such subjects as quality criteria, eutrophication, etc., contained in other chapters, may serve to underscore the role of estuarine water quality considerations in the overall problem of water quality management.

CONSIDERATIONS
OF AQUATIC LIFE

Protection of aquatic life from adverse water qualities is far more complicated in estuarine waters than in the streams which discharge into them, both because of the diversity of life itself and of the shifting and subtle environmental relationships in estuarine equilibria. For example, freshwater itself may be a pollutant in an estuary. An estimated loss of 90 percent of the oyster crop in Mobile Bay occurred in 1963 when unusual rainfall inland dropped the salinity of the bay below 10,000 mg/liter for a prolonged period at a critical time.

Although, as noted in Chapter 13, considerable experimental work has been done on the effect of toxic materials on fishes, much of this relates to standard test species or to the most valuable of freshwater species exposed to a single type of material. In the estuary, however, where numerous species may exist, and where all quality factors from all upstream beneficial uses are mingled, the data are difficult to evaluate. Each species of fish has its own period of life when it will be most susceptible to toxic compounds in the aquatic environment. In general, the younger fish are the most sensitive. Moreover, when several different toxic agents are present, strange effects can result. One agent may either enhance or decrease the toxic activity of another agent. This might be particularly true in saline waters where comparatively large concentrations of many ions are already present. In general, however, there seems to be a tendency for most poisons to be less toxic to fish in a brackish or marine environment than in a freshwater environment. This may result from either the adaptation of species to ions or to an antagonism between ions, or both. On the other hand, it may be more apparent than real since it is based on mortality rather than productivity observations. In this connection Butler [1] states:

Although acute toxic levels of natural and man-made pollutants are quite simple to demonstrate, the effects of low levels of toxicants are much more obscure, and disastrous changes in productivity levels might occur without significant mortality. Our ability to delineate the individual factors that are deleterious to estuarine forms will determine in large measure our success in preserving this part of the marine environment, which is so sensitive to pollution.

Fortunately many saltwater fish do not propagate in regions readily accessible to waste-water outfalls, but to some extent this only makes it more important to know the life history of each species of fish in determining the effects of waste discharges on fish life. An example may be drawn from the ecologies [2, 3] of some saltwater species found in the waters of California, Table 7-1.

Table 7-1   *Ecology of some saltwater species found in California waters*

Type of species	Species	Remarks: Possible effect of water quality in estuary
Freshwater spawners	Steelhead trout Striped bass King salmon Perches	Minimal effect as passage through bay (e.g., San Francisco Bay) is made by adult or near-adult fish.
Estuarine spawners (San Francisco Bay)	Herring	Spawn in spring in shallow flats, could be very susceptible to adverse quality factors.
	English sole Sand dab Starry flounder	Flatfish, bottom dwellers may be particularly affected by benthic conditions not characteristic of overlying water. English sole observed to develop growth on heads when dwelling in vicinity of sewer outfalls in San Francisco Bay.
	Bay smelt Sturgeon	Response to benthic and water quality conditions uncertain.
Estuarine residents	Perches Striped bass	Presence in bay as adults observed to increase when sewage pollution reduced.
Migratory from remote ocean centers	Jack mackerel Bonito Tuna	Practically unaffected by estuarine or near-coast water quality conditions.

To some extent the mobility of fishes enables them to escape to a limited degree from some adverse water quality factors. Shellfish, however, cannot escape from or avoid an adverse climate. Oysters are essentially obliged to live in estuarine waters since they seem to prefer brackish water in the 10 to 30 gm/liter range of salinity. Some species of clams are also estuarine dwellers. While adaptation to a wide range of salinity is characteristic of such shellfish, they are sensitive to many other quality factors in both gross and subtle ways. For example, oysters show considerable tolerance to low dissolved oxygen (D.O.) levels, but growth and fattening may be inhibited and shells deformed when D.O. sinks to 3 or 4 mg/liter for prolonged periods of time. As yet unknown factors affect their response to carbohydrates and temperature. From the viewpoint of aquatic biologists and conservationists the "maintenance of habitat" is becoming an increasingly important objective of water quality control. It is more and more being reflected in public policy, and scientific parameters for judging it are being refined. The latter are based on the important ecological rule [4] that "if the normal climate of an ecological system is changed, the number of species will decrease while the number of individuals will increase." This can be brought about by the addition of a waste water of somewhat different quality aspects to a natural water more or less in a state of equilibrium, without the addition of nutrients or a toxic agent. Such refinement of objectives and subtlety of response must inevitably suggest that increasing knowledge of the environmental responses of aquatic life to water quality factors will further intensify rather than relax the constraints which quality needs of estuarine waters may impose upon freshwater quality management routines.

PUBLIC HEALTH FACTORS

Concern for the effects of pollution on aquatic life involves questions of human health as well as of the welfare of aquatic species. Shellfish are typically filter feeders. Some species are capable of pumping up to 50 liters of water per hour, hence concentration of organisms characterizes their life process. Coliform organisms found in shellfish, on an MPN per 100 g basis, are from one to ten times the MPN per 100 ml of water in which they are immersed. These organisms may persist in the shellfish for 15 to 60 days after removal from polluted water. McKee [5] correctly suggests that "... shellfish constitute one of the weakest links in our defense against enteric infection, whether bacterial or viral." Further, a committee of the American Public Health Association stated in 1963 [6] "... there is ample evidence that failure to conform to coliform standards for drinking waters and shellfish harvesting waters has produced disease outbreaks." For this reason the quality of water entering an estuary from which shellfish are harvested is a matter of great concern.

No particular public health problem has been observed with fish except during canning. However, fish can become contaminated with coliform organisms within 1 hour of exposure and require 6 or 7 days to rid themselves of these organisms when transferred to clean water.

Health and its closely related factor of aesthetics are a consideration in the recreational use of estuarine waters. Epidemiological evidence to support water quality requirements for bathing beaches, water skiing areas, etc., is essentially nonexistent. Nevertheless, the level of general sanitary standards expected by the public, plus the problem of engendering a high degree of sewage treatment, if not applied "across the board," generates requirements for estuarine waters that are reflected in upstream quality management objectives.

ECONOMIC FACTORS

Of the several estuarine resource values having economic implications, fisheries and recreation are the two most concerned with water quality factors. Commercial fishing in California is not particularly an activity in estuarine waters. Prior to 1900, however, the 435-mile² area comprising San Francisco Bay accounted for 93 percent of the state's commercial fishing production—oysters, clams, mussels, crabs, and various species of fish [7]. After 1900, fishing pressure, siltation, and pollution led to a rapid decline in commercial activity. The contribution of water quality alone to this decline is impossible to assess with accuracy. Nevertheless, it is known that reduction of sewage pollution after 1950 restored an important amount of aquatic life, and the area became dedicated to sport fishing instead of commercial fishing.

The economic value of sport fishing in California is difficult to determine. Angling licenses in 1963 numbered 1.7 million [7]. In 1961, 20 percent of the 1.5 million license holders for that year fished in San Francisco Bay and the connecting delta, 15 percent of the total fishing days being spent in that area. Angling in the Bay, the delta, and the ocean is estimated [7] to represent 45 percent, with 55 percent being in freshwater. The number of California anglers was estimated in 1965 as 1.17 million with an annual expenditure of more than 107 million dollars. A sizable, if unidentifiable, portion of this total is concerned with waters where quality of influent water is a factor, largely in the preservation of aquatic environments suitable for fish life.

Pleasure boating, water skiing, aesthetic enjoyment, and other recreational activities involving large financial investment depend upon estuarine waters. In California alone there are estimated to be about 79,000 pleasure boats in operation (1962). However, the quality considerations of other estuarine resources are overriding and it seems unlikely that the public would tolerate a system of water quality management in which the needs of these activities would be the critical ones.

Unlike California, a significant portion of the commercial fisheries of other seaboard states is done in estuarine waters. The oyster industry of the East, South, and Pacific Northwest depends upon maintenance of quality of such waters, as do other segments of the seafood industry. In British Columbia, 62 percent of the commercial catch of salmon is made in estuarine waters. Its landed value was placed at 50 million dollars in 1963.

Industrial use of estuarine water is an important economic factor. Since this is mainly used for cooling purposes, temperature may become an environmental factor in sectors close to industrial return outfalls. Turning again to San Francisco Bay as an example, the 1962–63 average rate of industrial recirculation of water was 100 mgd [3]. During that same year more than 81,000 acre-ft of bay water was evaporated to produce industrial salts.

GENERAL CONCLUSION

Although the direct interchange of quality factors between the estuarine and freshwater sectors of the overall water resource is minimal, the resource values of estuarine waters impose limitations upon the quality of influent water. Therefore any system of freshwater quality management must take into consideration the quality requirements of this link in the transport chain. Furthermore, population concentration around major U.S. estuaries, public awareness of unsatisfactory conditions so close to home, and the subtleties of aquatic environment relationships tend to increase the pressure for greater upstream quality control.

REFERENCES

1 Philip A. Butler: "Reaction of Estuarine Mollusks to Some Environmental Factors," *Biological Problems in Water Pollution, Third Seminar, August 13-17, 1962,* U.S. Public Health Service Publication 999-WP-25, 1965.

2 E. A. Pearson: " 'Reduced Area' Investigation of San Francisco Bay," report to California State Water Quality Control Board, July 31, 1958.

3 "Report on Collation, Evaluation, and Presentation of Scientific and Technical Data Relative to the Marine Disposal of Liquid Wastes," California State Water Quality Control Board, 1964.

4 R. S. Fischer, A. S. Corbet, and C. B. Williams: "The Relationship between the Number of Species and the Number of Individuals in a Random Sample of an Animal Population," *J. Animal Ecology,* 12(42), 1943.

5 J. E. McKee and H. W. Wolf: "Water Quality Criteria," 2d ed., California State Water Quality Control Board, Publication No. 3-A, 1963.

6 "Coliform Standards for Recreational Waters," Progress Report, Public Health Activities Committee, Sanitary Engineering Division, *J. Sanit. Eng. Div., ASCE,* SA4, August, 1963.

7 "California and Use of the Ocean," Planning Study of Marine Resources prepared for California State Office of Planning, University of California, Institute of Marine Resources, IMR Ref. 65-21, October, 1965.

CHAPTER 8

Summary and evaluation of quality interchanges

INTRODUCTION

The quality interchange system under both natural and man-modified conditions has been presented in Figures 3-2 and 3-4. In subsequent chapters the relative magnitude, rate of growth, and quality aspects of untreated wastes have been indicated for the principal beneficial uses—domestic, industrial, and agricultural—which involve withdrawal of water from the surface- and ground-water resource. Some comment has likewise been made on the significance of these individual factors in water quality management. A full appraisal of the problem of operating the overall man-modified system for resource management objectives, however, requires consideration of several other aspects as well. Among the most important are:

1. The intensity of beneficial use of the water resource
2. The nature of quality factors introduced by beneficial use
3. The factors which cause variations in water quality
4. The natural quality changes that occur in water
5. The emergence of new factors that affect water quality
6. The quality requirements of beneficial uses
7. The resolving power and effectiveness of water quality criteria
8. The ability of present or foreseeable technology to upgrade the quality of water
9. The water quality objectives which society demands

An evaluation of some of these aspects may be made on the basis of concepts and data already presented. The others are worthy of individual consideration, as in the chapters which follow.

EXTENT OF USE OF FRESHWATER RESOURCE

The relationship of use to water supply is relatively easy to observe or to estimate in an isolated individual situation. However, if a national policy

74

of water quality is to be established, or regional planning for integrated and coordinated multiple beneficial use is to be practiced, a broader and less easily evaluated system must be considered.

In such a system, quality management may involve a combination of redistribution of the water supply, treatment of return flows, recycling and reuse of water, segregation of used and fresh waters, process changes to conserve water, limitation of quality objectives on individual sectors of the water resource, and a variety of other devices. To a large degree the appropriate combination at any point in time will depend upon the percentage of available supply that is put to beneficial use. Here the question arises as to what constitutes "available" supply. A difference in breadth of viewpoint in this matter leads one reporter to state with assurance that "we are running out of water," whereas another is equally confident that "there is no shortage of water—only maldistribution." In all probability the truth lies somewhere between these two extremes. While the citizen is confused, the engineer and economist seeks to identify the balance point between supply, demand, quality restoration, and redistribution beyond which degradation rather than generation of economic growth sets in.

The extent of utilization of the freshwater resource as estimated by Wolman [1] in 1962 is shown in Figure 8-1. From this figure the values shown in Table 8-1 are computed.

Table 8-1   *Summary of disposition*
of rainfall on land area of the United States

Disposition	Percent annual precipitation	Percent concentrated supply
Concentrated in streams and groundwater	29	100
Consumed loss in beneficial use	2	7
Streamflow not withdrawn	22	76
Streamflow not withdrawn, plus return flows	27	93
Return flows from beneficial use	5	17
Loss through noneconomic vegetation	32	

A cursory review of Table 8-1 would suggest that some 76 percent of the freshwater resource from streams and groundwater serves the following quality-related functions:

1. Beneficial uses not involving withdrawals
2. Dilution of treated and untreated return flows from beneficial use involving withdrawals

The question at once arises as to whether or not a greater fraction of this streamflow could be made available for the major uses of withdrawn water—

domestic, industrial, and agricultural. A second question is what would be the effect of greater withdrawals on the quality of the water resource, both in regard to (*a*) its suitability for other beneficial uses and (*b*) the technology needed to maintain the quality required by public policy. The answer to (*a*)

Figure 8-1   *Average distribution of precipitation in the continental United States. (Wolman [1]; reproduced by permission of the National Academy of Sciences–National Research Council.)*

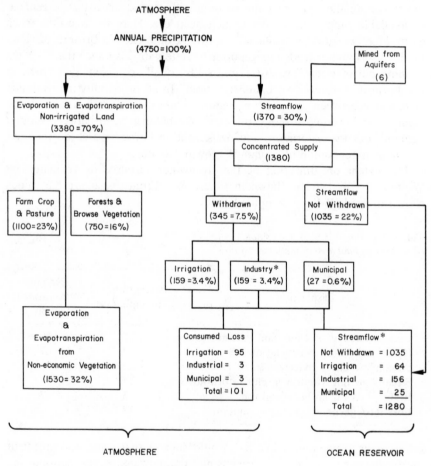

Legend and Values:

All figures in units of million acre-ft   Total Precipitation $=1552\times10^{12}$ gal/year
   one acre-ft $=43{,}560$ ft^3 or
   one acre-ft $=326{,}700$ gal   Total Precipitation $=4200\times10^{9}$ gpd

*The same water may be reused at points spaced along a single stream.

relates to the question of our running out of water. To answer (b) is to resolve essentially the entire problem of water quality management.

Some statistics presented by the Department of Commerce gave [2] a none-too-optimistic answer to the first of the foregoing questions. Of an average 4300×10^9 gpd precipitation, some 1260×10^9 gpd was estimated to remain after evapotranspiration, two-thirds of which goes into the ocean as floodwaters during one-third of the year. By use of storage it was considered economically feasible to provide a dependable minimum flow of 630×10^9 gpd. Of this amount, it was estimated that one-half would be necessary for wildlife, navigation, etc., and one-half available for consumption. In contrast, the 1980 estimate of consumption was some 60 percent of the available.

It is from such an estimate of the demand for water within the present century in comparison with the amount that can economically be put to beneficial use that a crisis in water supply is predicted. In this context, economic feasibility refers both to the cost of interbasin transfers and the cost of removing the effects of beneficial use. Here again, one may stress the need for economic optimization techniques, as mentioned in Chapter 3, if quality management is to be a rational factor in resources management.

Analysis of the system depicted in Figure 8-1 likewise reveals the basis of the more optimistic conclusions of "plentiful water" made by some observers. Seasonal or not, they contend, streamflow can be stored in surface reservoirs and in groundwater basins in the orderly process of achieving the beneficial uses of flood control, power production, low flow augmentation, irrigation, recreation, and wildlife preservation. Thus a greater percentage of water can be withdrawn and beneficially used. Restoration of quality by such users, however, is an implicit assumption which has its base in many interesting and often undocumented points of view. At one extreme is the belief that the heritage of America entitles its citizens to water of "wilderness" quality in all its surface waters, that industries and cities are immoral despoilers of this heritage, and that the public need only to set water criteria at primitive levels and have the law on all offenders. Closely related is the conviction that "if we can put a man on the moon we can surely overcome water pollution," that an opulent society can buy with dollars anything it desires, and that "science" is capable of finding a way to achieve any technology and to put it into practice with an accompanying expansion of the economy.

At the other extreme are beneficial users of water who find it hard to believe that they can "afford" to erase the effects of water use, that the advocate of "primitive" water quality will not revert to primitive economic concepts at bond-voting time, and that all sectors of the water resource must be dedicated to the highest possible beneficial use of water.

Nevertheless, increasing the percentage of the water supply going into beneficial use is one of the obvious ways to increase the productivity of water suggested by a consideration of Figure 8-1. Another is the reduction of noneconomic vegetation in order to increase the groundwater supply. Still others include recycling within any given beneficial use; sequential rather than parallel use of water by cities, industry, and agriculture; reduction of consumed loss, particularly in the stored water; and change of process in industry or development of new crop types in agriculture. All of these are, of course, under various degrees of investigation or practice. Their effects on the quality of water, and hence on water quality management techniques, are similarly in various stages of resolution. Some are discussed in the chapters which follow.

There is no doubt, however, that the intensity with which it becomes necessary to use the freshwater resource is one of the major determinants in water quality management methods and objectives; and conversely, our ability to manage water quality, technologically or economically, or both, may well limit the percentage of the freshwater resource that can be used, even if no consideration were given to its relationship to other resources.

NATURE OF QUALITY FACTORS INTERCHANGED

In the matter of water quality management it is necessary to establish a base-line concept. Specifically, a management system might deal exclusively with those quality factors which are altered by beneficial use, or it might seek also to control the interchange that would occur in nature if nothing was added or subtracted from water by man's use of it. Figure 8-2 and

Figure 8-2   *Natural interchange system*

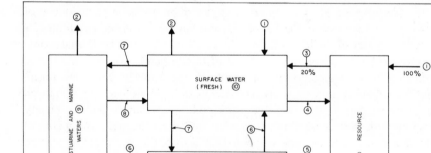

NOTE: THE CIRCLED NUMBERS REFER TO THE BOLDFACE NUMBERS IN TABLE 8-20.
    %   INDICATES FRACTION OF AVERAGE UNITED STATES RAINFALL OF 30 in. [3]

Table 8-2, which represent an elaboration of Figure 3-2, summarize the quality interchanges which might be expected to occur in a primitive or wilderness situation, unmodified by man. In this connection it is important to remember that construction of physical works such as dams, revetments, canals, etc., can interrupt and alter such a regime without the addition of any quality factor through beneficial use. (See Chapter 9.) Too often the engineer in designing such works seems to be unaware that in nature all things are in equilibrium, however tenuous or dynamic, and that

Table 8-2   *Summary of quality interchanges in natural system*

Symbol (Fig. 8-2)	Description	Principal quality factors
1	Meteorological water precipitation	Dissolved gases (CO_2, N_2, O_2, CO, SO_2), dust particles, smoke particles, bacteria, salt nuclides, dissolved vapors
2	Evaporation, evapotranspiration, volcanic activity, forest fires, wind pickup	Water vapor, salt nuclides, vapors from vegetation, organic and inorganic dust particles, bacteria, smoke particles, gases of combustion
3	Surface runoff	Silt, organic debris, silica, mineral residues of earth materials, soluble and particulate products of biodegradation of organic matter, bacteria, dissolved gases, soil particles
4	Floodwaters	Silt and other soil materials
5	Infiltration to groundwater	Dissolved minerals from surface debris and primary rocks, dissolved gases (CO_2, O_2)
6	Groundwater overflow (springs)	Mineralized water, intruded saline water
7	Infiltration to groundwater	Dissolved minerals from surface wash and groundwater outcrop; dissolved gases and compounds from organic life and decay; dissolved gases (CO_2, O_2)
8	Tidal waters	Increased salinity, dissolved minerals
9	Estuarine and marine water	(See Chapter 7)
10	Surface water	(See Chapter 9)
11	Groundwater	(See Chapter 16)

Table 8-3   *Summary of quality interchanges in man-modified systems*

Symbol (Fig. 8-3)	Description	Principal quality factors
1	Meteorological water precipitation	Dissolved gases (CO_2, O_2, N_2, NO_2, CO, NH_3, SO_2, H_2S, etc.); dust particles, smoke particles, bacteria, radionuclides, dissolved vapors.
2	Irrigation water supply	Same as **12**, freshwater resource.
3	Irrigation return water (leaching, overflow)	Same as **2** but with salts, nutrients, pesticides, organic debris, salinity increased by consumptive loss.
4	Domestic water supply	Raw water quality: same as **12**, suspended solids, bacteria, some dissolved solids and gases, removed by treatment processes (coagulation, filtration, softening, chlorination, etc.).
5	Domestic return water (waste water)	Raw water quality: same as **4** after treatment, plus degradable organic matter (human body wastes, ground garbage, grease, etc.); approximately 300 mg/liter added dissolved solids; bacteria; viruses; some industrial wastes **8**. Treated water quality: approximately 90 percent reduction in solids, bacteria, and oxygen-demanding properties added by domestic use.

the interruption of any part of the regime triggers an immediate search for a new equilibrium. Thus revetments have swept away beaches or filled harbors with sand; dams have been undermined by pickup of the riverbed by tailwater; and in a recent incident, the concentration of steelhead at the entrance to a fishway led to an almost total kill of other species unable to survive the appetite of steelheads in sufficient numbers to produce a significant spawn. Nevertheless, Figure 8-2 identifies the bridges at which the quality factors listed in Table 8-2 might be interrupted for quality management objectives, without regard to the consequences of interruption.

For the most part it is the man-modified system described in Figure 8-3 and Table 8-3 that enters significantly in quality management considerations. Leaving for later chapters a discussion of the methods of altering

Table 8-3 *Summary of quality
interchanges in man-modified systems (Continued)*

Symbol (Fig. 8-3)	Description	Principal quality factors
6	Stream discharge (surface and ground water overflow) to ocean	Same as **12** plus **3**, **5**, and **8** and partial self-purification.
7	Industrial water supply	Raw water quality: same as **12**. Treated as required by individual industry.
8	Industrial return water (process and cooling)	Same as **7** plus added organic matter, metal ions, chemical residues, etc., not removed by specific treatment. Increased temperature and salts concentration.
9	Industrial cooling water	Same as **14**.
10	Industrial cooling water, return	Same as **9** but with increased temperature and salts concentration.
11	Consumed water: Evaporation Evapotranspiration	Salt nuclides, dissolved vapors. (Principal effect is to increase concentration of salts in return waters and hence in resource pool **12**.)
12	Freshwater resource (surface and ground)	Quality governed by quality of inputs, **1, 3, 5, 8**, modified by natural systems and factors. (See Chapters 9, 10, 11, 13, 16, 17, 18.)
13	Atmosphere	(See **2**, Table 8-2.)
14	Saline water resource	(See Chapters 7 and 19.)
15	Tidal waters, intruded saline waters	Increased salinity and dissolved minerals.

quality at the various interchange points, or bridges, suffice it to emphasize here the nature of the quality factors which can be attacked at these points.

In order to give scale to the problem of quality management at the interchanges, estimates of the amount of water withdrawn in 1960 and predicted for 1980, together with the fate of withdrawals discussed in Chapters 4, 5, and 6, are summarized in Figure 8-3.

VARIATIONS IN WATER QUALITY

The quality aspects which have been related to meteorological waters, and to the effects of domestic, industrial, and agricultural use on water returned to the resource pool, are by no means constant in time. Neither are the quality attributes of the surface and groundwater resources themselves, although the latter is least affected by diurnal, seasonal, or annual aboveground variations.

Variations in quality occur in water from the same type of source as a result of a number of conditions, of which those shown in Table 8-3 are among the most significant [4, 5].

Figure 8-3   *Man-modified quality interchange system. Total withdrawals $= 224 \times 10^9$ gpd, 1960; 597×10^9 gpd. 1980.*

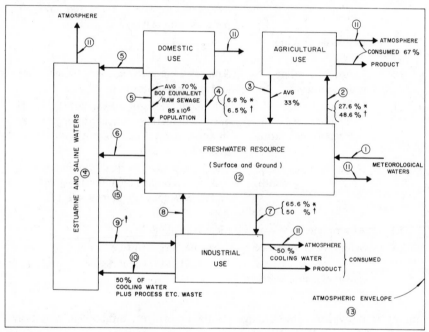

The factors listed in Table 8-4 bear upon the problems in a number of ways. They call attention particularly to the variations implicit in the interchanges summarized in Tables 8-2 and 8-3 and hence to the considerations which must be taken into account in the design of engineered subsystems on which quality management depends. Again, they suggest the interdependence of systems of water resource management and those involved in the management of other natural resources.

Table 8-4 *Factors in water quality variation*

Climatic conditions	Runoff from snowmelt—muddy, soft, high bacterial count. Runoff during drought—high mineral content, hard, groundwater characteristics. Runoff during floods—less bacteria than snowmelt, may be muddy (depending upon other factors listed below).
Geographic conditions	Steep headwater runoff differs from lower valley areas in ground cover, gradients, transporting power, etc.
Geologic conditions	Clay soils produce mud. Organic soils or swamps produce color. Cultivated land yields silt, fertilizers, herbicides, and insecticides. Fractured or fissured rocks may permit silt, bacteria, etc., to move with groundwater. Mineral content dependent upon geologic formations.
Season of year	Fall runoff carries dead vegetation—color, taste, organic extractives, bacteria. Dry season yields dissolved salts. Irrigation return water, in growing season only. Cannery wastes seasonal. Aquatic organisms seasonal. Overturn of lakes and reservoirs seasonal. Floods generally seasonal. Dry period, low flows, seasonal.
Resource management practices	Agricultural soils and other denuded soils are productive of sediments, etc. (See third item under *Geologic conditions.*) Forested land and swamp land yield organic debris. Overgrazed or denuded land subject to erosion. Continuous or batch discharge of industrial wastes alters shock loads. Inplant management of waste streams governs nature of waste.
Diurnal variation	Production of oxygen by planktonic algae varies from day to night. D.O. in water varies in some fashion. Raw sewage flow variable within 24-hour period; treated sewage variation less pronounced. Industrial wastes variable—process wastes during productive shift; different material during washdown and cleanup.

REFERENCES

1   A. Wolman: "Water Resources: A Report to the Committee on National Resources of the National Academy of Sciences–National Research Council," Publication 1000-B, Washington, D.C., 1962.

2   Anon.: *Water Works Eng.*, August, 1960.

3   George E. Symons: "The Hydrologic Cycle," *Water Sewage Works,* 100(7), July, 1953.

4   *Water Quality and Treatment,* AWWA Manual, American Water Works Association, New York, 1950.

5   E. F. Eldridge: "Return Irrigation Water–Characteristics and Effects," U.S. Public Health Service, Region IX, May 1, 1960.

CHAPTER 9

Quality changes
in surface and ground waters

INTRODUCTION

For the purpose of examining the inputs and identifying the factors which each might contribute to the quality of the overall resource, it is sufficient to lump all aspects of surface and ground waters together as though the freshwater resource pool were a common tank into which everything is discharged and thoroughly mixed. As we come to examine the quality changes which take place in the resource pool itself through natural phenomena, however, it is necessary to abandon this simple and comfortable fiction and to separate the various components of the freshwater resource for individual examination. Later on, when the subjects of "standards" and "quality requirements of beneficial use" are considered, even this breakdown will not be enough and it will be necessary to examine both the magnitude and the nature of inputs and of the water resource on a local basis. Finally, it will be seen that the selection and design of engineered subsystems, by which quality is managed for either local or national objectives, are governed by the parameters of each situation.

SUMMARY OF INPUTS

The inputs in terms of quality-controlling factors that contributed to the freshwater resource pool by meteorological waters, domestic use, industrial use, agricultural use, and other consumptive use of water have been summarized in Tables 8-2 and 8-3. For the purpose of examining the quality changes in surface and ground waters it may be useful to restate these same factors in the more detailed manner of Table 9-1.

QUALITY CHANGES
IN SURFACE WATERS

Flowing streams

In order to understand in even an elementary fashion the quality changes normal to flowing water, either with or without return waters from man's

Table 9-1   *Summary of quality inputs to surface and ground waters*

Contributing factor	Principal quality input to surface waters
Meteorological water	Dissolved gases native to atmosphere
	Soluble gases from man's industrial activities
	Particulate matter from industrial stacks, dust, and radioactive particles
	Material washed from surface of earth, e.g.:
	Organic matter such as leaves, grass, and other vegetation in all stages of biodegradation
	Bacteria associated with surface debris (including intestinal organisms)
	Clay, silt, and other mineral particles
	Organic extractives from decaying vegetation
	Insecticide and herbicide residues
Domestic use (exclusive of industrial)	Undecomposed organic matter, such as garbage ground to sewer, grease, etc.
	Partially degraded organic matter such as raw wastes from human bodies
	Combination of above two after biodegradation to various degrees of sewage treatment
	Bacteria (including pathogens), viruses, worm eggs
	Grit from soil washings, eggshells, ground bone, etc.
	Miscellaneous organic solids, e.g., paper, rags, plastics, and synthetic materials
	Detergents
Industrial use	Biodegradable organic matter having a wide range of oxygen demand
	Inorganic solids, mineral residues
	Chemical residues ranging from simple acids and alkalies to those of highly complex molecular structure
	Metal ions
Agricultural use	Increased concentration of salts and ions
	Fertilizer residues
	Insecticide and herbicide residues
	Silt and soil particles
	Organic debris, e.g., crop residues
Consumptive use (all sources)	Increased concentration of suspended and dissolved solids by loss of water to atmosphere

Table 9-1 *Summary of quality
inputs to surface and ground waters* (*Continued*)

Contributing factor	Principal quality input to groundwaters

Meteorological water	Gases, including O_2 and CO_2, N_2, H_2S, and H
	 Dissolved minerals, e.g.:
	     Bicarbonates and sulfates of Ca and Mg dissolved from earth minerals
	     Nitrates and chlorides of Ca, Mg, Na, and K dissolved from soil and organic decay residues
	     Soluble iron, Mn, and F salts

Domestic use
   (principally via
   septic tank systems
   and seepage from
   polluted surface
   waters)
	Detergents
	 Nitrates, sulfates, and other residues of organic decay
	 Salts and ions dissolved in the public water supply
	 Soluble organic compounds

Industrial use
   (not much direct
   disposal to soil)
	Soluble salts from seepage of surface waters containing industrial wastes

Agricultural use	Concentrated salts normal to water applied to land Other materials as per meteorological waters

Land disposal of
   solid wastes (not
   properly installed)
	Hardness-producing leachings from ashes
	 Soluble chemical and gaseous products or organic decay

NOTE: This list includes the types of things that may come from any contributing factor. Not all are present in each specific instance.

activities, it is necessary to have a concept of a stream as a living thing. From a relatively broad, and at the same time short, point of view as measured by geologic time, a stream in its natural setting may be said to be in dynamic equilibrium with other elements of the environment. Nevertheless, a complex set of transients makes up that equilibrium. At one season it gathers strength directly from meteorological waters, which rush in with a burden of soil and organic material stripped from the surface of the earth. At that time it tears at its banks, snatches up deposits from its own bed, grinds stones against one another, and generally destroys its former state of equilibrium. Then as precipitation ceases, the stream grows weary, laying down its burden somewhat in the same areas it recently ripped up. Too weak to carry a load, it finally takes on the quality characteristics of the groundwater on which it now depends for life.

Although the new stream bed may appear the same as before, there is a difference; the old degraded deposits have been replaced with new fertilizer elements and a crop of higher energy organic matter, hence the foundation is maintained for continuance of the biological food chain in the stream.

With never-ending variety the stream itself is a succession of environments replicated many times. In the rapids it picks up oxygen and carries it into more quiet pools which act as both sedimentation basins and life-sustaining systems. Here the nutrients and CO_2 released by bacteria (mostly aerobic), carrying out the biodegradation of organic sediments, support the microscopic plants which serve as food for small animal life, in a dog-eat-dog society that ranges from bacteria to the fishes and other higher forms. Even in the shallows, the downstream zone of each stone harbors sediments in which life abounds.

To the biologist unconcerned with the destiny of individual species in the food chain this may seem an idyllic situation. The philosophical attitudes at other levels of the food chain have not been reported. Nevertheless it is well understood that interruption of the food chain at any level is catastrophic to all forms above that level. Eliminate, for example, the nutrients and the water becomes a biological desert; destroy the fly larvae, etc., and fish life disappears.

The whole matter of stream pollution hinges on this factor of the maintenance of an uninterrupted food chain. However, the concern of this chapter is with the way a stream reacts in terms of water quality to classes of materials without particular regard to their origin, although in the interest of an overall understanding there is no reason not to identify various sources.

Reaction of flowing water to seasonal precipitation under wilderness conditions has been implied in the foregoing paragraphs. Fundamentally it cleans out its bed and replaces old deposits with other freshly mined from the earth minerals and from newly dead vegetation. In this there is a good balance in which the organic load does not reach such proportions that anaerobic decomposition of that load sets in. Neither does the load of sediment become inimical to the equilibrium, chiefly because it comes at a time when the stream has the physical strength to manage the load. This does not mean that individuals in the aquatic society do not live dangerously—the elements are no kinder to them than they are to humans. But it does mean that the society goes on after the flood with renewed vigor, just as man's does.

Let us now consider the reaction of a stream, in both a physical and a biological sense, to the introduction of other materials from other causes.

1. Excessive amounts of suspended matter

> (*a*) *Scientific agriculture* The very act of plowing up the natural sod and so uncovering large areas of the land surface has had

the effect of vastly increasing the sedimentary load introduced to streams by meteorological waters. Nothing new is introduced here as far as the stream is concerned. It accepts the burden and lays it down when it grows weary. The difference is that this time there is an bundance of fine material which has at least two major effects: It hastens the buildup of elevation of the riverbed and so increases its aggression on its own banks; and it obliterates much of the littoral environment in which fish spawn and their food chain flourishes.

(b) *Irrigation* Irrigation practice which results in runoff or over-flow may likewise introduce sediments eroded from the surface of agricultural soil. In this case a burden of fines is introduced to the stream at a time when its transporting power is at its lowest value. Consequently, fine material that would ordinarily be carried to the ocean is deposited in the pools and on the little bars behind stones, choking out the habitat of many organisms. Likewise, the constant turbidity over perhaps 128 days of irrigation cuts out much of the light utilized by plants which supply both food and oxygen to the water.

The total amount of silt produced has been estimated by Geier [1] at from 0.03 to 3.0 acre-ft/mile^2/year. Probably 1 ft in 4,000 years is his best guess. This amounts to about 7×10^6 ft^3/year from a 1,000-mile^2 drainage area, and represents the contribution from row agriculture, highway construction, building site clearance, etc.

(c) *Industrial activity* Certain industries, such as quarrying, mining, stonecutting, and construction, contribute sediments at times of low streamflow with the same effects as irrigation return water.

2. *Insufficient suspended matter* The wilderness stream, as has been noted, picks up its old sediments when it finds the strength but drops new ones in their place as it later weakens. Thus a migration of sediments from the source to the mouth of a stream is characteristic of flowing water.

Ponding of water behind dams, a feature of man's use of water for industry, agriculture, municipalities, flood control, recreation, and all other beneficial uses, has the effect of interrupting this intermittently advancing bed load of the stream. The result is that sediments originally laid down as high flows declined are picked up by the first high flow following the construction of the dam, but they are not replaced in amount. Hence the cycle of deposit and resuspension is phased out as the old load advances to the sea. Thus the stream reduces its gradient and an entirely new equilibrium of a different type is approached. It has a profound effect on the original stream ecology and the results are not always predictable. The lesson here, which man never has to cease learning, is that no simple problem in water resources management can be solved by a single solution. Any interruption of a dynamic equilibrium results in a shift toward a new equilibrium in which some of the elements in the old may be eliminated.

3. *Excessive organic loading*   Organic loading which introduces a demand for oxygen (BOD) in excess of the dissolved oxygen (D.O.) available in the stream will lead to anaerobic conditions such as depicted by the cycle of growth and anaerobic decay (Figure 2-3). Such a situation would represent an extreme case of excessive organic loading of a stream. An interruption of the food chain, however, can occur by a serious depression of the D.O. below levels (usually taken as 5 or 6 mg/liter) needed by higher forms of aquatic life.

The originally overloaded stream behaves somewhat as shown in Figure 9-1. It abandons aerobic decomposition of organic matter in favor of anaerobic processes. At some reach in the stream, however, anaerobic organisms run out of substrate, and aerobic organisms finish the job. As natural reaeration provides the oxygen, diffusion of oxygen into the flowing water is encouraged by the lowered oxygen tension; and later the nutrients from the stabilized organic matter support a

Figure 9-1   *Reaction of a stream to excessive organic loading. (Wisconsin Commission on Water Pollution.)*

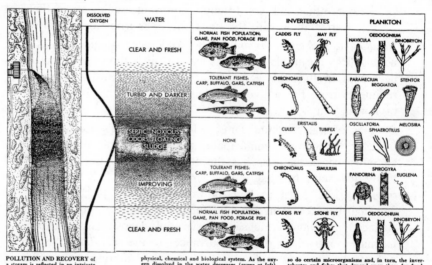

POLLUTION AND RECOVERY of a stream is reflected in an intricate physical, chemical and biological system. As the oxygen dissolved in the water decreases (*curve at left*), so do certain microorganisms and, in turn, the invertebrates and fishes that depend upon them for food.

richer aquatic society than the stream originally held. This society includes the green algae which also produce abundant oxygen.

Leaving to a later chapter a full discussion of the implications of a stream as a waste treatment process, it is sufficient to note here that a stream can recover from excessive organic loading but it must do so at the expense of quality characteristics acceptable to man, who wishes to use the water for other purposes; and at the expense of aquatic life which depends upon free (dissolved) oxygen.

This is the reason why BOD has been so much a factor in stream pollution-control discussions and measures. If the beneficial effects on aquatic life of the nutrients in sewage could be attained without the catastrophe of oxygen depletion, it is almost axiomatic that wildlife biologists would not object to raw sewage being discharged to the surface waters, even though health and other water use considerations would continue to require waste treatment.

4. *Miscellaneous factors* Reactions of a stream to various individual wastes may have the effect of changing water quality. Added nutrients increase aquatic life and the tastes and odors which may result. Toxic ions and compounds reduce aquatic life, but the river itself has no reaction to this factor. Various ions and materials may be precipitated in a stream or adsorbed on colloids in the season when sediments are transported most abundantly. Tannery wastes at low pH may turn brown at the higher pH of surface waters. In general, such miscellaneous factors are not profoundly involved in changes of quality due to flow of water.

Stored water

Concerning the self-purification of natural waters, Professor Sedgwick, one of America's first and greatest aquatic biologists, once said [2], "It is not so true that 'running water,' as that 'quiet water' purifies itself. We may even go so far as to say that the first requirement for the natural establishment of purity in surface waters is quiescence; but quiescence in rivers is ordinarily impossible." As noted in the preceding section, oxygen picked up in the rapids is utilized in the pools of streams. Pools, however, are far short of reservoirs in their depth, detention, and other characteristics, hence the quality aspects of stored water are worthy of separate consideration.

Changes in the quality of water resulting from the phenomena associated with impounded water take place in natural lakes and ponds as well as in artificial impoundments. To some degree these changes are intensified or extended by water resource management techniques. Reference has already been made (Chapter 6) to the increase in salts and total solids resulting from evaporation. Some 35 percent of the storage available for irrigation was estimated as lost by evaporation in 1960. Evaporation losses throughout the United States average 3 ft, and in the Southwest the value may be as much as 10 ft. Thus it may be said that stored water increases in salinity to a marked degree.

In recent studies of high-temperature water discharged to a reservoir, the reaction was to route the lighter water to the top where evaporation cooled the surface and no temperature change in the reservoir occurred. A somewhat similar gravity separation occurs with silt discharges which commonly sink to the depths and channel through a reservoir with little effect on the quality of water withdrawn for beneficial use, for pertinent reasons, from a higher elevation in the reservoir.

A decrease in bacterial content of stored water has been observed and attributed to lack of proper food, effects of sedimentation, disinfecting action of sunlight, depredations of other organisms, and devitalization.

The principal phenomena having an impact on the quality of naturally or artificially impounded water are those of stratification and vertical mixing. Most lakes, ponds, and reservoirs having any appreciable depth are subject to a season-related cycle of these two conditions. Since the sequence is indeed cyclical, a description of its nature must begin at some arbitrary point as in Figure 9-2. Here the familiar situation in a deep reservoir is depicted in late summer prior to a mixing of the contents of the reservoir commonly known as the "fall overturn."

As shown in the figure the reservoir water is stratified into three distinct zones. Approximately 10 ft of the top is subject to diurnal mixing due to wind disturbance and to sinking of a thin layer of surface water cooled, and consequently increased in density, by the ambient air temperature at

Figure 9-2   *Conditions in a stratified reservoir prior to fall overturn.*

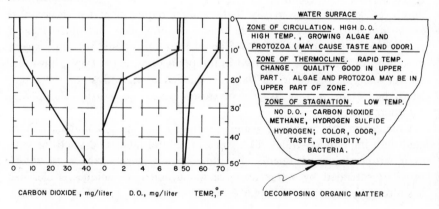

WATER SURFACE

ZONE OF CIRCULATION. HIGH D.O. HIGH TEMP., GROWING ALGAE AND PROTOZOA (MAY CAUSE TASTE AND ODOR)

ZONE OF THERMOCLINE. RAPID TEMP. CHANGE. QUALITY GOOD IN UPPER PART. ALGAE AND PROTOZOA MAY BE IN UPPER PART OF ZONE.

ZONE OF STAGNATION. LOW TEMP. NO D.O., CARBON DIOXIDE METHANE, HYDROGEN SULFIDE HYDROGEN; COLOR, ODOR, TASTE, TURBIDITY BACTERIA.

CARBON DIOXIDE, mg/liter   D.O., mg/liter   TEMP.° F   DECOMPOSING ORGANIC MATTER night. Thus dissolved oxygen is plentiful for aquatic life, as are the light and CO_2 on which plant life depends.

The zone of thermocline, probably caused by occasional deeper mixing due to thermal extremes of short duration, is principally characterized by a decline in temperature and oxygen content. From the standpoint of both temperature and the absence of the products of normal life and decay cycles, the highest quality of water exists in this zone.

In the zone of stagnant water, dead organic matter deposited by sedimentation exists in a water devoid of oxygen and at a temperature at which biodegradation may proceed, with all its attendant products.

Stratified water as illustrated in Figure 9-2 develops various quality characteristics. Living plankton in the zone of circulation give off essential oil in their life processes. Oil globules are likewise released when organisms die. These oils cause tastes and odors which are accentuated by the high water temperature. In excess numbers, plankton may cause turbidity and color, as well as tastes and odors, and may render the water unfit for recreational purposes.

In the stagnant zone anaerobically decomposing organic matter (dead leaves, surface wash, dead algae, and protozoa) produce organic acids and organic extractives. Extractives are productive of color; gases may be odorous; acids may produce tastes, odor, and color; and fine organic debris results in turbidity.

The fall overturn occurs when the surface water temperature approaches that of maximum water density (39.3°F), thus creating a condition of instability. Wind disturbance soon upsets the balance; and as the heavy surface water subsides, the entire reservoir goes into vertical circulation, giving the water mass the quality characteristics to be expected of such a mixture.

In relation to beneficial use, this natural cycle degrades the quality of the water resource and may well govern the nature of treatment works required to make the supply acceptable to the user, even though the situation exists throughout only a few weeks of the year.

Other quality-degrading phenomena may persist over longer periods of time. For example, wave action and seiches induced by winds stir up bottom deposits in the near-shore shallows, producing turbid waters which may then be carried out into the pond by circulation or by ice floes. Such turbidity also involves microscopic organisms and bacteria, as the littoral zone is the most active biologically.

In the absence of pollutants arising from man's activity, the quality changes naturally occurring in stored water are beneficial even though at the time of overturn the quality is lessened. Gases are released from the water mass and the products of anaerobic decomposition moved into an

Figure 9-3 *Summary of quality of natural waters.*

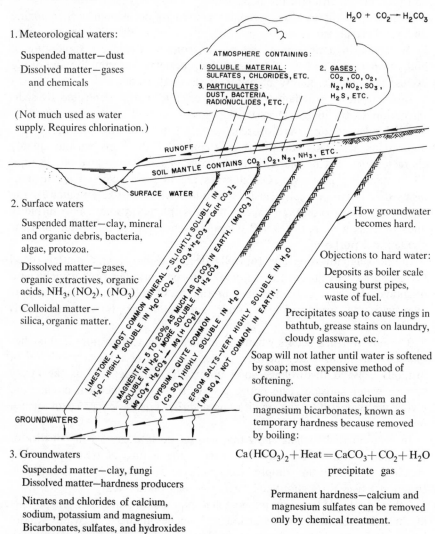

$$H_2O + CO_2 \rightarrow H_2CO_3$$

1. Meteorological waters:

 Suspended matter—dust
 Dissolved matter—gases
 and chemicals

 (Not much used as water
 supply. Requires chlorination.)

ATMOSPHERE CONTAINING:

1. SOLUBLE MATERIAL:
SULFATES, CHLORIDES, ETC.
2. GASES:
CO_2, CO, O_2, N_2, NO_2, SO_3, H_2S, ETC.
3. PARTICULATES:
DUST, BACTERIA, RADIONUCLIDES, ETC.

RUNOFF

SOIL MANTLE CONTAINS CO_2, O_2, N_2, NH_3, ETC.

SURFACE WATER

2. Surface waters

 Suspended matter—clay, mineral
 and organic debris, bacteria,
 algae, protozoa.

 Dissolved matter—gases,
 organic extractives, organic
 acids, NH_3, (NO_2), (NO_3).

 Colloidal matter—
 silica, organic matter.

LIMESTONE – MOST COMMON MINERAL – SLIGHTLY SOLUBLE IN H_2O + CO_2. $CaCO_3$ + H_2CO_3 $\rightarrow$ $Ca(HCO_3)_2$ SOLUBLE IN H_2O.

MAGNESITE – 5 TO 20% AS MUCH AS $CaCO_3$ IN EARTH. $MgCO_3$ + H_2CO_3 $\rightarrow$ $Mg(HCO_3)_2$ MORE SOLUBLE IN H_2O.

GYPSUM – QUITE COMMON $(CaSO_4)$ HIGHLY SOLUBLE IN H_2O.

EPSOM SALTS–VERY HIGHLY SOLUBLE IN H_2O $(MgSO_4)$ NOT COMMON IN EARTH.

GROUNDWATERS

How groundwater becomes hard.

Objections to hard water:

 Deposits as boiler scale
 causing burst pipes,
 waste of fuel.

 Precipitates soap to cause rings in
 bathtub, grease stains on laundry,
 cloudy glassware, etc.

 Soap will not lather until water is softened
 by soap; most expensive method of
 softening.

 Groundwater contains calcium and
 magnesium bicarbonates, known as
 temporary hardness because removed
 by boiling:

3. Groundwaters

 Suspended matter—clay, fungi
 Dissolved matter—hardness producers

 Nitrates and chlorides of calcium,
 sodium, potassium and magnesium.
 Bicarbonates, sulfates, and hydroxides
 of iron (stain porcelain fixtures, discolor laundry).

 Fluorides (stain teeth)

$$Ca(HCO_3)_2 + Heat = CaCO_3 + CO_2 + H_2O$$
$$\text{precipitate} \quad \text{gas}$$

 Permanent hardness—calcium and
 magnesium sulfates can be removed
 only by chemical treatment.

 Gases—Carbon dioxide, oxygen, nitrogen, hydrogen sulfide, and hydrogen.

 Surface-water treatment—filtration to remove color, odor, turbidity, bacteria;
 chlorination to kill bacteria.

 Groundwater treatment—softening, chlorination.

aerobic environment where stabilization can be completed. Thus the cyclical change in quality represents a cyclical purging of stored water.

Discharges from domestic, industrial, and agricultural use of water can, of course, drastically alter the natural regime. By overloading the reservoir with organic matter the annual overturn is incapable of purging the water mass and a long-term decline of water quality sets in, unless the reservoir is small enough to be scavenged by flowthrough of freshwater. In artificial reservoirs, organic debris left in the topsoil on the surface of the reservoir site may become the source of tastes and odors that persists for decades.

QUALITY CHANGES
IN GROUNDWATERS

The principal quality inputs to groundwater from meteorological water and from return flows from beneficial use are summarized in Table 9-1. In this case, water infiltering the soil from streams, ponds, canals, etc., is considered as meteorological, although during dry weather flow the surface loss to groundwater in one area may represent groundwater which previously outcropped further upstream. A comparison of the inputs to surface waters and groundwaters in Table 9-1, together with a consideration of the traditional clarity and purity of groundwaters, suggests at once that the soil mantle of the earth effects important changes in the quality of infiltering water. A full discussion of this phenomenon is left for Chapter 16, and attention is directed to the changes in water quality that occur as groundwater moves from its area of infiltration to its point of outcrop. Fundamentally, the change is one of increase in mineralization.

The principal natural quality changes in underground waters are depicted graphically in Figure 9-3. Here the nature of mineral pickup is likewise summarized, along with the objections to the resulting change in quality, and the quality control processes which man generally applies. For purposes of comparison a similar summary of the quality factors and control processes is included for surface waters as well.

REFERENCES

1 J. C. Geier: "Water Quality Concepts and Low-Flow Augmentation," *Proceedings of Invitational Seminar and Advanced Water Resources Topics,* University of Texas, April, 1964.

2 *Water Quality and Treatment,* AWWA Manual, American Water Works Association, New York, 1950.

CHAPTER 10

Refractory compounds as a factor in water quality

INTRODUCTION

Any quality factor which is resistant to natural or engineered treatment processes may be said to be refractory to such processes. Thus a whole spectrum of materials, produced by the chemical industry, ranging from common salt to exotic organic compounds, may be classed as refractory. The constraints on beneficial use of water resulting from the presence of refractory common ions picked up by water in contact with earth minerals or with decomposing organic matter are well enough known [1] to permit the establishment of criteria and standards for the protection of such beneficial uses. The same is true, to a somewhat less certain degree, of salts of heavy metals such as lead, chromium, nickel, copper, etc., associated with industrial waste waters. Scientifically feasible methods for the removal of most of these refractory materials are known, although the need for removal has not yet become sufficiently compelling to produce the technology or to force acceptance of the economic realities involved. Of particular interest in this group of compounds are the common nitrates and phosphates which serve as fertilizers in the natural cycle of growth and decay. (See Chapter 2.)

Because man has some understanding of the effects of these more ordinary substances in water, and has established an equilibrium, however uneasy, in their management, most current concern for refractory compounds as a factor in water quality is related to synthetic organic chemicals, particularly detergents and pesticides. Consequently, this chapter is directed to a consideration of the water quality management problems associated with these two.

SYNTHETIC DETERGENTS

The example of synthetic detergents is particularly significant in the history of refractory compounds, both because of the nature of public dissatisfaction with alkylbenzenesulfonate (ABS) and the method adopted to overcome it. The story of ABS is well documented in the literature and need only be summarized here as it relates to past and future problems of water quality management. It may be recalled that the use of synthetic household detergents began in about 1940 and within the next 25 years increased to some 4 billion lb/year. Of this total, some 75 percent of the active ingredient was ABS. Frothing of sewage effluents and of waters receiving return flows from domestic use become an increasingly common occurrence during the 1940-1965 period, and early investigation revealed ABS to be one of the major causes because it was refractory to conventional treatment processes.

The implications of detergent residuals in the freshwater resource became the object of many investigations and experiments to evaluate ABS in terms of its effect on human health, other biological systems, and water-treatment processes. For example, researchers at the University of California [2] and elsewhere demonstrated that ABS was resistant to conventional sewage treatment processes. The U.S. Public Health Service found [3] that acclimated biologically active sand systems were capable of degrading ABS. This was confirmed by the University of California in studies with finer soils and septic-tank percolation fields [4], and in field studies by Caltech [5]. Many researchers [6] observed the effect of ABS on vegetation, aquatic life, and higher animals. Ultimately it was concluded that while ABS under certain conditions might be harmful to certain aquatic organisms, the major objection was one of simple aesthetics—foaming of detergent-bearing surface and ground waters.

Although experimental work [7] had shown that ABS removal was technically feasible, it was refractory to conventional water and waste-water treatment processes; and the public saw no logic in expensive modification of treatment works which were meeting water quality management requirements based on BOD and solids reduction in domestic sewage. Consequently, there arose a very strong public sentiment for establishing by legislation a policy prohibiting the manufacture and sale of ABS.

Industry's response to the situation was an intensive search for a biodegradable substitute for ABS and, finally, an unprecedented changeover of the entire industry to a new product—LAS—in June 1965. Not only was this the first example of such an industry-wide action in the interests of water quality management, but also it was the first instance in which the ultimate disposability of the product was a major objective of the product designer.

From the standpoint of water quality management the example has important implications. It may well presage an era in which quality of the freshwater resource will be as serious a determinant of industrial processes as it is of plant location at the present time.

Results of studies [8, 9] of the biodegradability of the new surfactants LAS and alcohol sulfate in comparison with ABS are summarized in Table 10-1. They show that the normal secondary treatment processes used for domestic sewage are capable of a drastic reduction of the detergent present in such a waste water.

Table 10-1   *Removal of detergents by waste-treatment processes*

Process	Average percent removal		
	ABS	LAS	Alcohol sulfate
Primary sedimentation	2-3	2-3	
Septic tank	9.2	11.8	62.1
Septic tank + percolation field (normal)	73.9	97.4	99.6
Septic tank + percolation field (ponded)	54.5	97.1	99.7
Standard-rate oxidation pond	<40	93.1	98.0
High-rate oxidation pond	<15	56.2	95.2
Standard-rate trickling filter	35.0	84.7	
High-rate trickling filter	19.1	71.0	
Activated sludge	45-50	95	100

In die-away tests designed to reveal the expected behavior of surfactants in surface and ground waters, alcohol sulfate and LAS were about 90 percent degraded after one week, in comparison with 10 percent of ABS. At the end of 12 days the two "soft" detergents had degraded in excess of 95 percent, and ABS approximately 40 percent.

When considered in relation to water quality management the results of experiments with the new detergents indicate that detergent removal per se will not become an objective of waste-water treatment systems. Instead, the water quality objectives imposed on domestic return waters by reason of their oxygen demand, solids content, and biological nature should ensure that frothing of surfactant residue will not be a problem.

PROBLEM OF PESTICIDES

The question of the role of synthetic organic pesticides in the environment first came to widespread public attention with the publication of Rachel Carson's "Silent Spring" in 1962. Since that time savant and "true-believer" alike have repeatedly been heard from in public print. Although a full discussion, or even a fully documented expression of opinion or prejudice, is beyond the scope of this discussion, it might be well to give some scale and concept to the problem as it concerns the quality of our water resource.

In spite of alarmist reports, synthetic organic pesticides are not commonly used with reckless abandon even though their use is widespread. Nevertheless, the objectives of the use of pesticides have generally been limited to quite narrow concepts. That is, not enough thought has been given to the shift in equilibria which must necessarily follow the upsetting of an existing one. It is in this fact that a real danger of unknown consequences lies. Specifically, in rationalizing the use of synthetic pesticides, only two ecological streams seem to have been considered—man and his insect and botanical enemies [10]. The effects of introducing pesticides into other ecological streams which bear less direct, but eventually profound, effects on man is the great unknown factor which he can ill afford to ignore.

GENERAL NATURE OF PESTICIDES

Pesticides fall logically into two very broad classes—herbicides and insecticides. Each of these groups might be further broken down in countless ways, such as chemical nature, toxicity, persistence, solubility, movement in soil with percolating water, retention on soil colloids, volatility, breakdown products, and ultimate fate in the environment. In fact, serious research is continuously being directed to such factors. A recent publication of the Public Health Service [11] summarizes 437 important reports since 1944, nearly one-half of which have been published since 1960.

The foregoing factors, of course, bear upon the role of pesticides in water quality, as well as upon their more direct effect upon man. Toxicity of insecticides, however, is the factor given most attention, with the presumption that insecticides may appear in the water resource being the subject of both gross monitoring of receiving waters and of specific studies of the fate of insecticidal compounds in nature.

Most of the modern insecticides fall into two broad classes—the chlorinated hydrocarbons and the organic phosphates. The former includes such materials as DDT, BHC, dieldrin, endrin, heptachlor, toxaphene, TDE, aldrin, methoxychlor, and chlordane. Examples of the organic phosphates include malathion, parathion, TEPP, EPN, chlorthion, diazinon, guthion, dipterex, and demeton.

THE SCALE OF PRODUCTION AND USE

In 1960 there were more than 100 organic insecticides available. Estimates of production range from 500 to 1000×10^6 lb of organic pesticides, plus 1×10^6 lb of nonorganics (e.g., arsenicals, Cu and Zn sulfates, etc.) annually. Similar estimates of the acres of land to which pesticides were applied in

1959 range from 100×10^6, or about 5 percent of the United States, to 30×10^6 acres sprayed at least two times each year.

HOW PESTICIDAL
WATER POLLUTION
MIGHT OCCUR

There are several ways in which pesticides, particularly insecticides, might find their way into the freshwater resource:

1. By surface wash or runoff into streams
2. By movement with percolating water into groundwaters
3. By direct application to bodies of water for pest control
4. By drifting onto surface waters from adjoining treated areas
5. By return irrigation waters
6. By spills and wastes from pesticide manufacturing
7. From cannery and other food-processing waste waters

TOXICITY OF PESTICIDES

There is no doubt that pesticides of several varieties have extremely high indices of toxicity—that is the purpose for which they were created. Of the pesticides, the herbicides have given least concern in terms of water quality, although some can damage plants at a distance should they migrate through soil with percolating water or be carried into irrigation water by surface wash from treated land. Experience, however, would indicate that instances of such damage are not widespread. Evidence is cited [11] that 2,4-D, for example, is rapidly inactivated under soil conditions favoring activity of organisms. As a sodium salt, it has been observed to travel no more than 1 in. in soil with 1 in. of rainfall. TCA likewise may be expected to remain in the topsoil; 2,4-D in acid or ester form, however, is reported to be easily leached by water. Both 2,4-D and 2,4,5-T are known to be degraded by microflora in soils. Persistence of several herbicides has been shown to be inversely related to temperature and soil moisture, and directly related to the intensity of dosage. Furthermore, it varies with the type of soil.

By far the most important of the pesticides are the insecticides which in some cases have contaminated irrigation water long after the chemical has been applied. The chlorinated hydrocarbons listed in a preceding section are often thought to have the greatest potential for damaging the water resource because of their long residual activity. On the other hand, their low solubility in water is an offsetting factor, and hence the true significance of insecticides on agricultural soil or forest vegetation is difficult to assess in terms of water quality.

The organic phosphate insecticides are believed to be less dangerous to fish than the chlorinated hydrocarbons because they are less stable, some breaking down quickly in water.

Of the hydrocarbons, endrin appears to be the most toxic and TDE and BHC the least toxic to fishes. Of the phosphates, guthion seems by far the most toxic.

Lethal concentrations of most insecticides in various experiments have been summarized by Klein [12], and to a lesser extent in Chapter 13.

EVIDENCE OF DANGER
TO WATER QUALITY

Evidence that pesticides may endanger water quality particularly in relation to its suitability for aquatic life, and especially in the form of catastrophic incidents, is presented in many references. The following, compiled from several sources [12, 13, 14, 15, 16], is typical of the wealth of individual observations reported.

DDT in concentrations of 5 to 20 parts per billion (ppb) has been observed in the Detroit, Missouri, Mississippi, and Columbia Rivers. Aldrin at a concentration of 1 ppb was found in the Snake River. As early as 1950, fish were decimated in 15 streams tributary to the Tennessee River in Alabama by an insecticide applied over a wide area. In 1933 sanitary engineers from the Robert A. Taft Sanitary Engineering Center of the U.S. Public Health Service recovered DDT from the Detroit River and St. Clair Lake in both raw and treated water. DDT was present in the Detroit River and persisted in samples taken over a 6-month period.

In 1955, in St. Lucie County, Florida, 2,000 acres of salt marsh were treated for the sandfly larvae by the application of dieldrin at a rate of 1 lb/acre. It is estimated that 20 to 30 tons of fish were killed in canals bisecting the marshes.

TDE used to control the Clear Lake gnat on a lake of the same name in California appears to have led to the death of hundreds of grebes. Samples of fat from one bird showed a high concentration of the chemical. Similarly, samples of visceral fat from several species of fish taken over a period of 6 to 8 months after the last spraying showed from 40 to 2,000 mg/liter of TDE.

Recent observations in widely separated incidents suggest the probability that DDT now pervades the entire aquatic environment of the earth. It has been found in local fishes 2,000 miles from land in the Pacific Ocean. The Sanitary Engineering Research Laboratory of the University of California has sought unsuccessfully to find cultures of *Chlorella* that are DDT free. Even pure cultures nurtured for hundreds of generations in laboratories are found to contain measurable quantities of DDT.

Tarzwell and Henderson [15] of the U.S. Public Health Service have described a runoff study conducted near Atlanta, Georgia. Here 4.66 lb of dieldrin were applied per acre of grassland to control white fringed beetle. Bioassays showed that the first runoff from rain was toxic to fathead minnows in dilutions of one to three. On the third rain, fish mortality was less than 50 percent. The concentration of dieldrin was estimated at 0.128 mg/liter. Evidently only a small fraction of the chemical was washed out by the rain. It is not known whether the remainder was lost by weathering or remained in the soil to be washed out by future rains.

Walton [15] has described a case of underground water pollution near Henderson, Colorado. Groundwater contaminated by arsenicals which eventually formed 2,4-D traveled some 3 miles in 8 years to affect crops. Seven years later some 60 miles^2 had been affected seriously.

At Montebello, California, seepage of 2,4-D from a manufacturing plant contaminated the city's water supply. The plant was shut down within 30 days but taste and odor persisted in the water for 5 years.

Various similar instances of water pollution by pesticides are to be found in the literature.

SIGNIFICANCE OF FINDINGS

Although the potential of water pollution by pesticides can be shown by theoretical considerations and actual incidents to be worthy of deep concern, it must be recognized that the known deleterious effects on aquatic life are quite small considering the quantity and extent of use of pesticides in the United States. It is therefore necessary to consider the relation between the risks and the benefits to be derived from the use of pesticides.

It is impossible to gainsay that modern agriculture's ability to feed an opulent society would certainly fail if it were necessary to return to the "man with the hoe" to control weeds, or to compete with worms for every apple or ear of corn. The historic intervention of sea gulls which saved the crops of early Mormon settlers at Salt Lake from grasshoppers was not repeated. Thereafter it was up to man himself. In California and elsewhere in America, weeds and insect pests are controlled by airplane dusting. The integrity of United States forests depends upon similar procedures. Necessity for such activity was emphasized by a representative of the University of California Agricultural Extension Service who estimated in 1963 that 90 percent of the food grown by man could be destroyed by pests if continual wars were not waged against them.

From the viewpoint of water quality management the question then arises as to how these realities of life are to be reconciled with the objectives of society relative to the national water resource. The answer would seem to lie along two parallel lines of research. The first is directed to a further understanding of the long-term effects of pesticides in various ecological chains in order that the risk to man as well as to the quality of his water resources may be accurately assessed. The possibility must be entertained that the price of plentiful food may be an occasional catastrophe to an aquatic population. At the same time, the possibility exists that the long-term effect may be a catastrophe which man cannot readily correct. To that end the second stream of research is directed to the production of time-unstable compounds that will do their appointed task and then decay prior to reaching the resource pool—much as in the case of detergents—and to the elimination of pests by methods of biological confusion which do not involve the use of toxicants.

REFERENCES

1 "Summary Report, Advanced Waste Treatment-1, June 1960-December 1961," Sec. TR W62-9. Robert A. Taft Sanitary Engineering Center, U.S. Public Health Service, May, 1962.

2 P. H. McGauhey and S. A. Klein: "Removal of ABS by Sewage Treatment," *Sewage Ind. Wastes*, 31(8), August, 1959.

3 G. G. Robeck: "Degradation of ABS and Other Organics in Unsaturated Soils," *J. Water Pollution Control Federation*, 35(10), 1963.

4 S. A. Klein, D. I. Jenkins, and P. H. McGauhey: "The Fate of ABS in Soils and Plants," *J. Water Pollution Control Federation*, 35, 1963.

5 J. E. McKee, *Final Report on Wastewater Reclamation at Whittier Narrows*, Report to California State Water Quality Control Board, September 29, 1965.

6 David Eckhoff, David Jenkins, and P. H. McGauhey: "Evaluation of Improved Type Detergents," University of California, Sanitary Engineering Research Laboratory Rep. 64-12, December, 1964.

7 S. A. Klein and P. H. McGauhey: "Detergent Removal by Surface Stripping," *J. Water Pollution Control Federation*, 35(1), January, 1963.

8 S. A. Klein and P. H. McGauhey: "Degradation of Biologically Soft Detergents by Wastewater Treatment Processes," *J. Water Pollution Control Federation*, 37(6), June, 1965.

9 P. H. McGauhey and S. A. Klein: "Degradable Pollutants—A Study of the New Detergents," Paper, *Third Conf., Intern. Assoc. on Water Pollution Research*, Munich, September, 1966.

10 F. E. Egler: "Pesticides in Our Ecosystem," *Am. Scientist*, 52(1), January, 1964.

11 "Pesticides in Soil and Water. An Annotated Bibliography," U.S. Public Health Service Publication 999-WP-17, July, 1964.

12   Louis Klein: *River Pollution. 2: Causes and Effects,* Butterworth & Co. (Publishers), Ltd., London, 1962.

13   *Proceedings, Conf. Physiological Aspects of Water Quality,* U.S. Public Health Service, September 8-9, 1960.

14   "Water Resources Activities in the United States: Pollution Abatement," Senate Select Committee Print No. 9, 86th Cong., 2d. Sess., January, 1960.

15   C. M. Tarzwell, C. Henderson, and G. Walton: *Proceedings, Nat. Conf. Water Pollution,* U.S. Public Health Service, December 12-14, 1960.

16   R. M. O'Donnell: "Review of the Effects of Pesticides upon Water Pollution," unpublished paper, University of California.

CHAPTER 11

Eutrophication:
the effect of nutrients

INTRODUCTION

In a previous chapter it was pointed out that under wilderness conditions mineral salts brought in from weathered soil, plus degradable organic matter of terrestrial and aquatic origin, furnish the nutrients which support the normal aquatic food chain. It was particularly noted that such materials seldom are present in sufficient concentrations to lead to anaerobic conditions. This is to say that the amount of mineral and organic matter is the limiting factor in the abundance of aquatic life. And whenever such factor is limiting, the system is nutrient sensitive and can be expected to react quickly to the addition of fertilizer compounds at a low energy level; or to change in population characteristics if the compounds enter at a high energy level. In terms of the aerobic cycle of growth and decay, it is the left-hand half of the cycle which reacts first to the introduction of stable nutrients, whereas the right-hand sector will be first to react to unstable organic materials, possibly going into the anaerobic cycle with catastrophe to the aquatic society.

Highly treated domestic and organic industrial wastes; return water from irrigated agriculture; and runoff from farmland, golf courses, lawns, and other land areas to which commercial fertilizer is applied are the principal outside sources of biochemically stable nutrients entering surface waters. Exerting little or no BOD, these compounds have the effect of enriching the aquatic life. Incompletely treated domestic and organic industrial waste waters, surface wash from vegetated land and from organic swamps, and dead aquatic organisms within the receiving water itself are the principal sources of unstable organic matter which must first be degraded through an oxygen-demanding process before they appear as nutrients to other than the microbial population of the water.

**EUTROPHICATION
OF LAKES AND RIVERS**

It is reported that the Swedish biologist, Naumann [1], in 1919 was first to designate nutrient-poor waters as "oligotrophic" and nutrient-rich waters as "eutrophic." Utilizing Naumann's concept, the term "eutrophication" may be used to describe the process of maturation of a lake from a nutrient-poor to a nutrient-rich body of water. The wilderness lake in its early years is almost universally oligotrophic, unless it is created in an organic swamp. If it is of appreciable size and depth so that the inflow-outflow relationships are small in comparison with the volume of the lake, many centuries may be required for the organic buildup to reach such proportions that the lake becomes nutrient-rich and continues to remain so through recycling of nutrients regardless of the relatively small annual influent of such materials. In other words, eutrophication, although the ultimate destiny of a lake, may be very slow in developing. If, however, artificial enrichment occurs as a result of human activities, lakes may become eutrophic within a very short period of time.

Thomas [2] notes that "In northern Switzerland . . . all lakes of 20 meters depth or more with at least 500,000 square meters of surface area belonged to oligotrophic types until a few decades ago. . . . Today most of these lakes are in a eutrophic state as a result of artificial fertilization."

An extensive list of lakes showing early eutrophy was published by Hasler in 1947 [3]. A partial summary of this list is presented in Table 11-1. It includes lakes ranging from 286.1 to 0.18 miles^2 in surface area and from 414 to 4.2 m in depth.

The advance of eutrophication of a lake receiving artificial fertilization is dramatically demonstrated by the Zurichsee in Switzerland. It consists of two basins—the Obersee and the Untersee—separated by a narrow passage. Hasler states that in the past 50 years "The Untersee, the deeper of the two basins, at one time a decidedly oligotrophic lake, has become strongly eutrophic, apparently owing to urban effluents originating from a group of small communities (the City of Zurich is located at the downstream end of the lake but contributes no sewage). The shallower of the two received no major urban drainage and has retained its oligotrophic characteristics."

**OBJECTIONS
TO EUTROPHICATION**

Implied, if not indeed explicit, in the foregoing paragraphs is the concept that eutrophication, like sin, is something to be abhorred by all right-minded water resource management people. What, in terms of water quality, do we have to lose by accelerating the natural process of nutrient buildup in ponded waters? To answer this question it is only necessary to consider the

Table 11-1 *Lakes showing early
eutrophy owing to domestic drainage*

Lake	Area, miles^2	Maximum depth, meters
Alpine lakes		
Zurichsee (Unter)	25.86	143
Zurichsee (Ober)	7.72	50
Hallwilersee	3.90	48
Rotsee	0.18	26
Zugersee	15.01	198
Murtenersee	8.81	46
Freifensee	0.33	35
Worthersee	7.50	85
Lago di Como	56.33	414
Finnish lakes		
Lohjanjarvi	47.10	56
Vesijarvi	56.37	40
Mommilanjarvi	1.46	100
Lehijarvi	286.10	19
England		
Windermere	5.72	67
United States		
Mendota (Wisconsin)	15.20	25.6
Wingra (Wisconsin)	0.31	4.2
Monona (Wisconsin)	5.44	22.5
Waubesa (Wisconsin)	3.18	11.1
Kegonsa (Wisconsin)	4.91	9.6
Geneva (Wisconsin)	8.53	43.2
Delavan (Wisconsin)	2.83	17.2
Nagawicka (Wisconsin)	1.43	28.8
Pewaukee (Wisconsin)	3.59	13.8
Lake Washington (Washington)	33.83	65.2

SOURCE: Hasler [3]. Reproduced by permission of the publisher.

quality characteristics of a lake water before and after it has become eutrophied.

Large lakes in their original or wilderness state have, as Thomas [2] explains, a very large volume of water in the hypolimnion (zone of stagnant water). Characteristically such lakes appear blue or greenish blue, the water is clear and transparent, and the phosphorus content of the water is so low that plants cannot make use of all the available nitrogen. "There is only a quantitatively moderate development of phytoplankton organisms even in the surface water, and littoral plant production is also slight. The high permeability of light prevents the formation of a distinct thermocline and provides the plants with good conditions for development even at great depths."

In such lakes the alkalinity and dissolved oxygen vary but little from top to bottom and there is no production of ammonia, hydrogen sulfide, or other gases of decomposition in the depths. Although some of these conditions are conducive to organic growth, nutrients are the limiting factor; hence the production of plants and phytoplankton is slight during the entire year. In contrast the deep fauna is comparatively rich, whitefish (*Coregonen*) being the predominant species. The fall overturn of such a lake would produce none of the adverse quality conditions discussed in Chapter 10.

It should be pointed out that small lakes, even in the original state, may be subject to the seasonal low-quality limitations described in Chapter 10. In them the ratio of water volume to influent organic matter—leaves, surface wash, pollen grains, insects, etc.—may be quite small, leading to anaerobic conditions in the depths and to quality degradation previously described in connection with the turnover of reservoirs.

The objections of eutrophication, however, are not limited to the degradation of the quality of stratified water. The nutrient substances present in sewage effluents frequently lead to the production of explosive "blooms" of algae of such intensity that clear sparkling lakes are transformed to turbid-colored bodies of water. Decomposition of the algae then produces obnoxious odors and floating decomposing mats of organic matter. Thomas [2] graphically describes the conditions in these terms: "Plankton algae make the water turbid and discolored with unpleasant shades of green, yellow, brown, and violet. The floating layers of plankton algae are particularly disagreeable to swimmers, boaters, fishermen, inhabitants of the banks and people walking nearby. Shore algae also are often unpleasant."

He describes the growth of diatoms on stones in eutrophied Swiss lakes, and the masses of *Oscillatoria* which overgrow the mud or lake chalk bottom of the shallows. Part of the oxygen from this growth is held back in the tangled mass of algal filaments. "These gas bubbles finally result in patches larger than hand size being detached from the bottom so that they float on the surface and form repulsive flat cakes there that look like the skin of a toad. This unpleasant phenomenon is overcome only by rain and wind in April or May. The original biotopes of the 'toad's skin' are then overgrown by other algae . . . which then rise to the surface of the lake in summer and disfigure the shores." He describes in detail the coating of stones in water 0 to 3 m deep with diatoms and algal filaments to a depth of several centimeters. Algal pads of *Cladophora* rise to the surface annoying everyone who somes in contact with them. In Lake Zurich such masses constrain the natural motion of reeds in the wind with the result that the reeds break at the surface of the restraining mat and themselves contribute to a mass of decomposing organic matter. The result is a situation disadvantageous to fishermen, aesthetically objectionable to the public, and conducive to gases of decomposition such as hydrogen sulfide which are in themselves objectionable.

The fish population of sewage-fertilized, i.e., eutrophied, lakes has been observed to change from whitefish to coarser species. In the United States, eutrophication of lakes has caused great concern. The lakes at Madison, Wisconsin, and Lake Washington at Seattle have probably been given widest publicity. More recently, national attention has been drawn to Lake Tahoe, one of the finest of the few remaining oligotrophic lakes in the world, where managing man's occupancy of the lake basin without creating the conditions which would lead to early eutrophication of the lake presents some unique problems.

Concerning the Wisconsin lakes, Rohlich [4] has noted that: "As early as 1882, algal blooms were reported as having occurred in Lake Mendota and Monona.... Over a period of more than twenty-five years argument and debate continued in regard to the odor nuisances, the fertilization of lakes, and the extensive algae growths."

In the case of Lake Washington it remained a relatively clear oligotrophic lake until quite recently. Anderson [5] noted that since 1950 "prominent changes have occurred, particularly in 1955 when for the first time there appeared an increased growth of phytoplankton made up mainly by the blue-green alga *Oscillatoria rubescens,* a notorious indicator of pollution in many lakes." He points out that since 1955 the annual crop of algae has increased with the result that with "the increase in productivity the transparency of the water has decreased and oxygen consumption and nutrient release in the hypolimnion (zone of stagnation) during summer stratification has progressively increased." The oxygen deficit has increased threefold in the depths since 1933, and in 1957 the deepest waters became anaerobic for a short period for the first time.

The consequences of enrichment of rivers are similar to those of lakes in their effects upon beneficial use. In the wilderness state sufficient nutrients are brought into the stream to support aquatic life without detracting from the appearance and other qualities of the water. In fact, a river in its original state is suitable for fishing, swimming, and development for public water supply. Enrichment of such a stream has little effect on the reaches where velocity is high. In the pools and quiescent zones, however, algae and higher aquatic plants soon develop. A thick growth of weeds depreciates the sport of fishing. Swimming is impossible. And tastes, odors, and other factors compound the problem of reclaiming it for water supply.

Exclusion of further nutrient discharges to a river will result in its reverting in time to its original state of nutrient-limited aquatic life. The lake, however, is essentially irreversible because of recycling of nutrients. Solar energy is the only input needed to keep the cycle going indefinitely, since the normal influx of fertilizer compounds is more than enough to offset the discharge losses. As previously noted, there is a slow accumulation of nutrients in a lake even under wilderness conditions, but "instant eutrophication" is the result of human activities.

FACTORS INFLUENCING EUTROPHICATION

Although the amount of nutrients is an influencing factor, there is no simple relationship between the maturing process of a lake and the amount of nutrients present in its waters. The rate at which eutrophication occurs is governed by a combination of factors related in a complex fashion which is not fully understood. Nevertheless, there is no doubt that urban drainage is one of the most critcal factors. Rohlich [4] has noted that: "Essentially all climatic, physical, and biological factors have an influence on the eutrophication process particularly as they relate to the distribution, availability, and utilization of the nutrients required by the algae in their metabolism." Rawson [6] stresses the complexity of the relationship of environmental factors which lead to eutrophication of lakes. Geology, geometry of the lake, temperature, dissolved oxygen, pH, calcium, iron, nitrogen, phosphorus, silica, and organic material seem all to be involved in a complex fashion of unknown nature. Sawyer and Lackey [7] and others, however, stress the predominance of nitrogen compounds and phosphorus as determinants of algal blooms. In a report authored by Sawyer and Lackey in 1943 [8], nitrogen and phosphorus were identified as the culprits in the eutrophication of the Madison lakes, the major source for Lake Mendota being the disposal plant for the City of Madison. The report further stated that "inorganic nitrogen and phosphorus were found to be critical factors in the productivity of the lakes, and that during the year 1942-1943, Lake Waubesa, the most heavily fertilized lake in the survey, at least 65 percent of the inorganic nitrogen and 89 percent of the inorganic phosphorus entering the lake were derived from nonagricultural drainage." Later studies by other observers in 1949 found the less frequent blooms in Lake Mendota, in comparison with its companion lakes, to be related to its lesser intake of nutrients. However, there are sufficient inputs from nonurban sources to produce algal blooms in the lake when other environmental factors are favorable.

Rohlich [4] calls attention to another case in which algal blooms are not triggered by the return waters from irrigation, industry, or cities. This is the upper Klamath Lake which for 60 years has developed unsightly and offensive smelling conditions each summer. "The precise explanation," Rohlich notes, "for the unusually high productivity is as yet not fully explained but it was noted that the drainage water from 92,000 acres of muck soil agricultural land which has been drained and diked, as well as drainage from

136,000 acres of essentially natural marsh land, contains humic leachates from the marsh soil, and analysis of the leachate shows the nitrogen content in summer to be more than double that in the winter. Algae culture studies using media enriched with humic water showed that the production, measured by weight of algae, was 70 percent greater in the humic-enriched cultures than in the unmodified controls."

Thomas [2] reported results of experiments in which nitrogen and phosphorus were added to samples of water from 46 different lakes, in the amounts of 20 mg/liter nitrogen and 2 mg/liter phosphorus. Over a 2-month period the plankton had greatly increased in all cases. The results demonstrated that "only nitrates and phosphates come into question as minimum substances in such lakes . . . the addition of nitrogen and phosphorus compounds to lakes and rivers is sufficient to stimulate some types of algae to greatly increased growth, which can however, have very detrimental results."

Maloney [9] of the U.S. Public Health Service attempted to determine the minimum concentrations of nutrients necessary to support the growth of algae. His tentative conclusions were that nitrogen should be below 0.1 mg/liter and phosphorus below 0.01 mg/liter, with essentially no iron present, if nutrient scarcity is to prevent algal growth. Similar conclusions have been reached by other observers [4].

It is generally conceded today that phosphorus is more important than nitrogen in causing algal blooms. In fact, an algal bloom may result from excess phosphate out of all scale with the normal ratio of nitrogen to phosphorus in plant growth. This is the result of the development of types of algae which can fix nitrogen from the atmosphere provided there is plenty of available phosphates.

Although in the normal cycle of nature nutrients have been shown to be the critical factors in eutrophication, it may not be concluded that removal of nitrogen and phosphorus is sufficient to maintain the oligotrophic characteristics of a given body of water. If small amounts of nitrogen and phosphorus are present, some missing growth factor may be the triggering mechanism. Magnesium, trace elements, or some entirely unknown growth factor may be the stimulant needed to induce eutrophication. There has been but little observation of growth stimulants because waste-water treatment practice has been directed to stabilizing rather than removing fertilizing compounds. Consequently, nutrients have been the overriding factor in eutrophication.

THE PROBLEM
OF QUALITY MANAGEMENT

The entire concept upon which treatment of domestic wastes has been predicated is the matter of BOD reduction. Standards, of course, specify a limit on suspended solids as well, but the sewage-treatment process has been built around satisfying the oxygen demand of degradable organic matter before releasing domestic waste water to the resource pool. This simply means that the right half of the aerobic cycle of growth and decay must be more nearly completed by the treatment process as the demand for higher and higher quality of effluent is imposed by regulatory agencies. Until quite recently, however, the prevention of eutrophication of receiving waters has not been considered as an objective of sewage treatment, nor has it been a goal of water quality regulation. Historically each time the water quality objectives have been tightened, the discharger of organic wastes was required to bring the state of nitrogen and phosphates in his effluent nearer to that of stable inorganic nitrates and phosphates. Within the past 2 or 3 years, however, pollution-control agencies notably in the eastern and southern United States, have begun to set receiving-water standards which simply cannot be met by extending the degree of oxidation of nitrogen and phosphorus in the discharge. It is not clear whether the regulatory agencies are aware of this, but the consulting engineer, confronted with a requirement that cannot be met because organisms are quite as happy with nitrate as with ammonia, is in a quandary. We have simply not yet come up with any technologically and economically feasible method of nitrogen removal, nor has nitrogen removal been generally recognized as an objective of waste treatment. Deionizing of water, distillation, anaerobic fermentation with denitrification, and other procedures can, of course, remove nitrogen. However, the parameters necessary to design engineering system, as well as the economics of such systems, have yet to be developed.

A similar problem exists in the case of phosphates. Phosphate removal methods were developed and published [10] more than 15 years ago. Currently, one installation based on the method of phosphate removal by alum precipitation is going into operation in the Lake Tahoe Basin on what is patently an experimental basis as far as operational parameters and economics are concerned. Although the technique of removing phosphate is known, some very difficult problems may arise if phosphate removal becomes an objective of waste-water treatment. To a large degree these relate to the sources of phosphorus in sewage.

Phosphorus finds its way into domestic wastes via a number of routes. Phosphates are concentrated in the leaves of growing plants; thus the household garbage grinder introduces a portion of the phosphate content of sewage. Milk contains an important amount of phosphorus, and the use of milk on a large scale in the diet of Americans makes an important contribution of this element to the waste waters. The builder in the more than 4 billion lbs of synthetic detergents sold each year in the United States contains phosphates. In fact, possibly 30 to 50 percent of the phosphorus content of sewage comes from household detergents. There are already indications that within a few years phosphates rather than detergents may cause public concern. This time, however, industry cannot respond in the fashion it did when the nuisance of ABS froth on water was abhorred by the public, by rearranging the carbon atoms in the molecule. There is no known way to formulate detergents without phosphates, nor does there appear to be any satisfactory answer beyond phosphate removal—a prospect which under current costs may well confront the public with some difficult choices along the road to its goal of "unpolluted," "pure," or "clean" water.

There are other ways than nutrient removal from sewage which might theoretically overcome the eutrophication of lakes but they do not seem of practical importance. Among the expedients sometimes suggested are harvesting of fish crops, removal of algal growths and plants from lakes, siphoning off deep water from lakes to remove the nutrients produced by biodegradation of organic matter, artificial aeration of deep water, and chemical poisoning of algae. While each of these techniques may have some role in the management of water quality, in the long run the control of eutrophication by engineered systems must either involve prevention of sewage discharges or nutrient removal from the waste water. Expedients might then be helpful in reestablishing oligotrophic conditions in certain instances; but in general, the control of aquatic growths must depend upon nutrient removal, and assay of the growth-stimulating ability of the remainder, before a waste water is returned to the freshwater resource pool.

LAKE TAHOE:
A CASE IN POINT

By the time the prevention of eutrophication became recognized as an important goal of water quality management, most of the lakes in the United States had passed from the oligotrophic to the eutrophic state. Consequently, the contributing factors and their critical levels can never be evaluated. Lakes slowly developed objectionable characteristics, after which it was possible only to catalog the gross causes and the objections to their effects. A major exception is Lake Tahoe, situated on the California-Nevada border, one of the few oligotrophic lakes in America.

Lake Tahoe [4] is one of the highest (6,200 ft above sea level) and deepest (maximum 1,645 ft, average 990 ft) lakes in the world. Geographically, it lies within a 4-hour drive of some 3 million people and within a day's drive of 10 percent of the population of the United States. The climate of the region makes possible year-round use of the lake as a recreation center. The lake water is unrivaled for clarity, having an extinction coefficient the same as distilled water and a Secchi disk transparency averaging 28 m. The waters are well mixed, with oxygen extending to great depths. The thermocline exists at 15 to 75 m and no annual overturn has been observed. Bottom temperatures are too low for biological activity to reduce the quality of water in the hypolimnion. Nitrogen and phosphorus levels are below 0.01 mg/liter, dissolved oxygen is about 9.0 mg/liter, and chlorides average some 1.7 mg/liter. The lake is considered to be nitrogen sensitive in contrast with the phosphorus sensitivity of most oligotrophic lakes. Thus Lake Tahoe may be said to be unique in that waters of a mineral quality acceptable for drinking purposes (see Chapter 12) would represent a considerable degree of pollution of the lake.

Clarity of water and the natural setting of Lake Tahoe give it an aesthetic appeal, the preservation of which has become the objective of many citizen and public groups. And prevention of eutrophication is widely recognized as the means of attaining this objective. Prevention, however, is not easily attainable. The reasons range from technical to economic and jurisdictional.

Factors which tend to presage eutrophication of Lake Tahoe are numerous. The lake, which is 21.6 miles long and 12 miles wide, has a shoreline of 71 miles, 90 percent privately owned. Population pressure against this shoreline is extremely great. For example, in 1962 the maximum summer population was 132,000 (11,400 minimum). By 1980 this figure is expected to reach 400,000. Food, fertilizer for lawns and golf courses, and the normal organic and inorganic debris associated with human living are imported to the basin. Export of nutrients is essentially zero except for the small amount carried out by overflow from the lake into the Truckee River. Sewage-treatment goals, as elsewhere, have long been either stabilization or underground disposal, in a closed geological basin. No harvesting of vegetation for export from the basin occurs. Thus there is assembling in the Lake Tahoe basin all the physical factors which make for eutrophication of the lake.

On the economic side year-round occupancy of the basin, together with some seasonal construction opportunity for profit, contribute to an attitude of preservation of local economy first and water quality second. On the jurisdictional side is the fact that 2 states, 5 counties, and more than 25 public agencies have authority in one facet or another of the area's activity. Thus in spite of a community of water quality goals, a time dimension of near-geological order characterizes antieutrophication measures.

From a strictly water quality management viewpoint several contributions to techniques may come from the Lake Tahoe example. Experiments in nutrient removal have been triggered by the local situation; export of treated effluent as a short-term, or perhaps permanent, expedient may well relax water rights constraints of long acceptance; and jurisdictional arrangements for areawide action may be developed. At the same time, scientific knowledge of great value in the control of eutrophication may result. Above all, the Lake Tahoe example underscores the very frail nature of water quality management techniques as the nation faces the need for managing this aspect of its water resource.

REFERENCES

1   E. Nauman: "Nagra Synspunkter Angaende Plankston Okologie," *Svenske. Bot. Tidskr.* 13, 1919.

2   E. A. Thomas: "The Eutrophication of Lakes and Rivers: Cause and Prevention," *Biological Problems in Water Pollution, Third Seminar, August 13-17, 1962,* U.S. Public Health Service publication 999-WP-25, 1965.

3   Arthur D. Hasler: "Eutrophication of Lakes by Domestic Drainage," *Ecology,* 28, 1947.

4   G. A. Rohlich: "Lake Tahoe Waste Disposal Study," Engineering Science Incorporated, Arcadia, California, 1963.

5   G. C. Anderson: "Recent Changes in the Trophic Nature of Lake Washington—A Review," *Trans. of the 1960 Seminar on Algae and Metropolitan Wastes,* U.S. Public Health Service Sec TR WG1-3, 1961.

6   D. S. Rawson: "Some Physical and Chemical Factors in the Metabolism of Lakes," *Problems of Lake Biology,* Publication 10 of the American Association for the Advancement of Science, 1939.

7   C. N. Sawyer and J. B. Lackey: "Investigation of the Odor Nuisance Occurring in the Madison Lakes Particularly Monona, Waubesa, Kegonsa from July 1942 to July 1943," *Report of the Governor's Committee,* 1943 (also 1944, 1945).

8   C. N. Sawyer and J. B. Lackey: "Plankton Productivity of Certain Southern Wisconsin Lakes as Related to Fertilization," *Sewage Works J.,* 17, 1945.

9   Thomas E. Maloney: Private communication to L. W. Weinberger, January 22, 1963.

10 W. L. Lea, G. A. Rohlich, and W. J. Katz: "Removal of Phosphates from Treated Sewage," *Sewage Ind. Wastes,* 26(3), 1954.

CHAPTER 12

Criteria and standards in water quality management

INTRODUCTION

Preceding chapters have identified the factors which change the quality of water, have related these factors to sources, and have noted the conditions under which they vary. It remains to consider the relation of these quality factors to the needs of various beneficial uses, and the means by which the quality of the water resource pool is controlled. Specifically, we are confronted with the necessity for quality control, both as regards the nature of return water that users shall be permitted to discharge, and the nature of water acceptable for any beneficial use. To do so, however, requires that we first describe quality in quantitative terms. It is not enough to say, for example, that turbidity is aesthetically objectionable. We are not going to have a resource pool of crystal clear water, so someone must decide how much turbidity the eye can detect and how much we are willing to pay to achieve such a limit.

But describing water quality precisely is not by itself enough. Control of water quality for any purpose, be it protection of the supply for some beneficial uses or the conditioning of process water, involves treatment. The goal of water treatment thus becomes that of altering or upgrading quality to a level appropriate to the intended use. Water treatment is accomplished by engineered systems; and the goal of sound engineering includes a concern for minimum cost consistent with health, safety, and product requirements. Therefore, if quality is to be changed, some way must be found to decide when the change has been carried far enough, in the interests of both economy and suitability of the finished water to its intended use. This way involves two concepts—"standards," or "requirements," and "criteria"—which should be clearly understood.

116

STANDARDS, REQUIREMENTS, AND CRITERIA

McKee and Wolf in their monumental work "Water Quality Criteria" [1] note that "The term 'standard' applies to any definite rule, principle, or measure *established by authority*. . . . The fact that a standard has been established by authority makes it quite rigid, official, or quasi-legal. An authoritative origin does not necessarily mean that the standard is fair, equitable, or based on sound scientific knowledge, for it may have been established somewhat arbitrarily on the basis of inadequate technical data tempered by a cautious factor of safety." The danger of such standards has been pointed out by Frank M. Stead of the California State Department of Public Health [2], who notes that "today's estimate is tomorrow's law." Nevertheless some sort of parameters are necessary. "We need a yardstick, even if it is made of rubber," continues Mr. Stead. To this end McKee and Wolf suggest that "A far better way to describe an administrative decision by a regulatory body is 'requirement.' It represents a requisite condition to fulfill a given mission. It does not necessarily have the connotation of scientific justification nor does it give an impression of immutability."

"A 'criterion,' " note McKee and Wolf, "designates a means by which anything is tried in forming a correct judgment respecting it." They note the need for applying numbers to criteria but warn against letting them ripen into rigid standards.

When water is drawn from the resource pool its quality in quantitative terms can readily be determined by physical, chemical, and biological analyses of varying precision. And, except in the case of human health and the welfare of other biota, the minimum acceptable change in each of these characteristics can be established through experimental procedures. This exception, however, is a serious one because our chief interest in this whole matter concerns ourselves and the flora and fauna in our own food chain. Therefore, our experiments get involved with epidemiology, which involves time and disaster; with statistics, which involves more of the same; and with biological responses, which get involved with adaptations and secondary responses that are difficult to evaluate. The values thus obtained can then be utilized as *criteria* for judging the suitability of the water for specific beneficial use. Assuming that the change necessary to make a water suitable for the intended use is attainable by known and economically feasible processes, the treatment system can be engineered. However, when used water is to be returned to the resource pool with the goal of protecting the interests of all beneficial uses, looking out for the public health, and keeping the quality of the pool such that reuse by all is within attainable limits, *standards* normally become the parameter used. These may be either stream standards or discharge standards, as discussed later.

VALUE OF STANDARDS

Standards which can be kept dynamic and flexible have certain recognizable values:

1. Measurement of quality factors is encouraged.
2. They permit self-control by dischargers.
3. They preserve fairness in application of police power.
4. They furnish a historical documented story of an event and thus assist in controlling the future.
5. They make possible the definition of a problem.
6. They establish goals for design of systems.
7. In setting standards one must face his own ignorance, or in facing arbitrary standards one must say definitely why they are inappropriate.
8. They assess what we are getting from a system.

WHERE STANDARDS ARE NEEDED

In Chapter 13 the standards of water quality currently used in relation to public water supply, agriculture, and industry are outlined. There are, however, several areas in which it is generally conceded that standards are yet needed. These include:

1. Chlorinated hydrocarbons in aquatic environments
2. Transparency of water
3. Nutrients in water
4. Safety in freshwater recreation
5. Safety of reclaimed water
6. Growth stimulants in water

CLASSES OF STANDARDS

There are four major classes of standards [2]:

1. Standard for etiological agent itself, e.g., pesticides.
2. Index standard, in cases where some associated factor is easier to measure than the agent itself, e.g., coliform organisms which measure potentiality for disease-producing organisms.
3. Precursor, or element that enters into a reaction which affects quality. BOD and nutrients are precursors, and in themselves not important agents.
4. Major environmental factor which produces some result worthy of concern, e.g., solar energy which drives biochemical reaction.

BASES OF PUBLIC POLICY

The adoption of standards involving numerical specifications of a spectrum of quality factors grew out of the goals of society as expressed in public policy. Protection of the public health was the first goal to be so implemented. Originally public health concern was directed to the prevention of specific illnesses. However, as time went on, aesthetic goals became associated wtih health, and standards for the prevention of nuisance or offense to the senses of sight and smell became common. The words "pure, wholesome, and potable" appeared in public health laws. They call for the expression of quality in numerical terms which can be interpreted as defining pure, wholesome, and potable—all noncontinuous variables which are ill-defined qualitative concepts.

But public health did not forever define the goals of water quality control. New agencies were created to deal with the quality of water where protection of the interests of beneficial uses not directly involving public health was concerned—aquatic life, recreation, industrial and agricultural supply, etc. These agencies, like the health agencies, must establish suitable criteria and standards.

Currently, a totally new goal of water quality has been added to that of protecting the public health and other beneficial uses. This is a social goal represented by organized and unorganized citizens who demand "clean" or "pure" water per se. How shall the demands of citizens be expressed in "standards" without turning history back to Indian days—an eventuality little advocated by the "pure water" enthusiasts—is a question for which the answer has not yet unfolded. But it does foretell the day when the return waters from beneficial uses are going to be required to look more and more like the natural water which might exist if the uses did not. Particularly do these "true believers" express a concern for aquatic life. In fact, McKee and Wolf's researches into the literature revealed that the number of references pertaining to the effects of pollutants on aquatic life far exceeds those relative to any other beneficial use.

To put this increasing social goal into perspective, Mr. Stead lists five zones of concern for the control of any aspect of the environment, including the water resource [2]. In descending order of urgency they are as follows:

1. Zone of aesthetic enjoyment (top zone)
2. Zone of physical comfort, at bottom of which man becomes aware of non-well-being
3. Zone of chronic illness or morbidity (chronic ill effect)—impaired performance of man or system
4. Zone of acute morbidity (ill effects acute and serious)
5. Zone of simple survival, at bottom of which is death or failure of a system (bottom zone)

The objectives of society may be to achieve the boundary between any two of these zones in any, or in all, cases. For example, the standards of water for public consumption have crept up from the bottom zone to zone 2, or in cases where the taste, odor, hardness and other uncomfortable factors are economically infeasible, to the top of zone 3.

A similar goal is developing for other beneficial uses as time goes on. And, as has been noted, the achievement of the zone of aesthetic enjoyment may well come to dominate the standards imposed on all beneficial uses of water—i.e., domestic, industrial, and agricultural return waters. Hence it is extremely important that standards be more like the "requirements" as previously defined: not permanent in magnitude for all time, but suited to the goals, the technology, and the economics of society at each stage of its development.

BASES ON WHICH STANDARDS ARE ESTABLISHED

At whatever level public policy is directed it can only establish the goals which are to be achieved by the adoption of standards, regulations, or criteria. Someone must eventually express quality factors in numbers which presumably will establish the desired boundary condition. One or more of the following bases may be used:

1. Established or going practice
2. Attainability, either easily or reasonably attainable
   (a) Technologically
   (b) Economically
3. Educated guess, making use of best information available
4. Experimentation (e.g., animal exposure)
5. Human exposure
   (a) Taking advantage of occurring catastrophe
   (b) Experimenting with humans directly
      (Note: The same concept applies to plants and animals, including aquatic life.)
6. Mathematical model or treatment (e.g., probability, mode, percentile, MPN coliforms)
      (Note: Biological events are geometric, hence single standards cannot be used to describe the phenomenon in any case.)

In the existing water quality standards detailed in Chapter 13, the educated guess and attainability have figured heavily.

A thorough study of the origins of standards of water quality reported by Professor McKee in 1952 [3] revealed the sources to be "technical personnel" or "a committee of representatives of the interested departments," either with or without the added obscuring factor of statewide hearings. In one case the frank reply to Professor McKee's questionnaire was that "we have no records or minutes to show the development of these standards and the former executive secretary-engineer who handled these matters is no longer with the Commission." More recently the Bureau of Sanitary Engineering of the California State Department of Public Health endeavored to trace to its origins the value of 500 mg/liter which appears as the recommended upper limit for total dissolved solids (TDS) in the Public Health Service Drinking Water Standards. While rational men might agree that it is a reasonable standard, the search revealed that the mind of man does not recall nor do his records reveal its origin. Most likely it represents a value widely attainable in the surface waters of an America much less populated than today. As 500 mg/liter is becoming increasingly unattainable in many cases, it is now important either to document its validity or to develop some new value to take its place. The former, as has been noted, is impossible; the latter is doubly difficult. To base the parameter on attainability one must first determine just what is attainable in a water resource pool subject to an ever-increasing recycling of waste waters; and he who would base it on some rational or experimental curve of TDS number versus water quality is doomed to frustration.

Attainability itself is by no means an easily identified parameter, except in instances of some specific constituent of water for which no scientifically feasible process of removal is known. Therefore, attainability has not always been accurately estimated and subsequent embarrassment has accrued to the standard-bearer. In one case where the presence of a lead-using industry suggested the need to set a standard for lead in the waters receiving the industry's waste discharges, only intelligent alertness on the part of the technical staff prevented an unhappy decision. In this case those charged with the responsibility of setting the standard suggested 0.1 mg/liter lead as a reasonable value for the receiving waters in view of what is known about lead poisoning and the beneficial uses to be protected. However, a survey revealed that the muds upstream from the industry naturally contained 10 to 250 mg/liter of lead. Eventually a more realistic and liberal standard was established on the grounds of attainability.

Perhaps the best researched parameter of water quality is the MPN of coliform organisms. Even it involves an element of the attainable, else why accept 1 organism per 100 ml when no organism at all is the goal. The answer, of course, lies in the epidemiological evidence that at this statistical level of contamination other vectors of intestinal diseases far outweigh the hazard of water. The parameter is therefore adequate under United States conditions but it is by no means exportable as a criterion of water quality to regions such as, for example, South Africa where the intensity of typhoid and cholera pollution of sewage is far greater than in the United States and where these diseases are not the only major health hazards of waste waters.

Avenues of approach to the problem of establishing a quantitative degree scale applicable to each of the many significant factors which affect the quality of water for each of a wide spectrum of uses, of course, transcend attainability and epidemiologically demonstrable catastrophe. Bioassay with 50 percent kill of test organisms must be translated into zero-kill concentrations. Results of tests on rats and dogs must be escalated to protect their larger cousin—man.* Social science must replace chemistry in defining "pure" water. And so on.

WATER QUALITY CONTROL

Once public policy has been enunciated and the bases on which water quality standards are to be established have been accepted, some method must be found for applying the underlying concepts to water quality control. Since it is the quality of the resource that is to be protected, such terms as "water quality criteria," "receiving water standards," and "pollution control standards" have been applied to the control measures.

Two general types of standards have been used as a basis for water pollution control:

1.   Those which set "stream standards" for receiving waters
2.   Those which set "effluent standards" for wastes discharged

Stream standards are of two classifications:

1.   Dilution requirements
2.   Standards of receiving water quality which depend on:
   (*a*)   Establishment of threshold values for various pollutants, or
   (*b*)   The beneficial use to which the water is to be put

Effluent standards are likewise of two classes:

1.   Those which restrict strength or amount of pollutants to be discharged (e.g., Delaware River Commission)
2.   Those which specify degree of treatment required

Standards of receiving water quality often involve a system of stream classification or zoning. This has the advantage over effluent standards in that the system takes into account dilution and assimilative capacity of the stream and hence makes for economy of treatment works for pollution abatement. On the other hand, standards are hard to define, and cumbersome and difficult to administer.

*The method here is illustrated by the following formula:

$$\frac{\text{Lifetime (2 yr) tolerance of healthy rat}}{10} = \text{lifetime tolerance of healthy man}$$

$$\frac{\text{Healthy man}}{10} = \text{frail man}$$

STANDARDS BASED
ON DILUTION

An example of stream standards based on dilution are those set by the British Royal Commission in 1912 for sewage and sewage effluents. These standards, which are summarized in Table 12-1, served satisfactorily for many years. They were developed to satisfy the requirements of small streams in populous areas receiving strong sewage (two to four times that of United States sewage).

Table 12-1 *Stream standards based on dilution*

Classification of standard	Required condition of sewage or effluent		Type of sewage treatment presumably satisfying the standards
	5-day, 65°F BOD, ppm	Suspended solids, ppm	
General standard	≦20	≦30	Complete treatment
Special standards— ratio of receiving water to sewage flow:			
150 to 300	. . .	60	Chemical precipitation
300 to 500	. . .	150	Plain sedimentation
Over 500	. . .	. . .	No treatment required

SOURCE: Imhoff and Fair [4]. Reproduced by permission of the publisher.

Although standards based on dilution are not suited to the United States conditions of today, they served the objectives of water quality management for many years under the conditions for which they were developed to a degree of satisfaction far greater than might be expected from the simplicity of the criteria involved.

STANDARDS BASED
ON STREAM CLASSIFICATION
AND TREATMENT

One of the first approaches in the United States to water quality management through stream standards is summarized in Table 12-2. This concept takes cognizance of the fact that the water resource is not a single pool, but consists of a number of sectors which require individual consideration by reason of difference in intrinsic character and the beneficial use requirements. In this case, receiving waters are divided roughly into four classes according to the highest use to which the stream is to be dedicated, and the criteria of minimum acceptable quality set forth. Such a system is typical of many early state laws established at a time when domestic sewage was the principal offender. Consequently, the standard went on to include the required treatment. This ultimately made the law all but impossible to administer. Particularly when industrial wastes were involved, the regulatory agency often had little information on how the return flows should be treated in order to achieve the specified quality goal. Hence no discharger could be held legally responsible for failure of his methods if they conformed to required treatment requirements yet failed to meet the stream standards.

STANDARDS BASED ON STREAM CLASSIFICATION

The shortcomings of criteria or standards based on stream classification and specified waste treatment were recognized in later legislation which set similar water quality criteria but made no mention of the methods for their attainment. Those recommended by the Interstate Commission of the Potomac River Basin, and by the Water Pollution Control Boards of states such as New York, Pennsylvania, South Carolina, West Virginia, etc., are typical examples of this type of criterion.

Table 12-3 derived from the Potomac River Commission data is illustrative of receiving water quality criteria which set minimum stream requirements for arbitrary classes without attempting to dictate the treatment measures by which they should be achieved.

These criteria, it is explained by the commission, are to be used as a guide, and only then in connection with a sanitary survey. Pollution agencies are to regulate waste discharge so that these standards are maintained. No rigid standards are set because of the wide variations in natural river conditions.

A stream standard such as summarized in Table 12-3, however, is not without its own limitations. Two are especially important from the standpoint of water quality management. First, stream classification tends to dedicate a receiving water to a particular level of use for an overlong period. For example, a single major industry may persist on a stream dedicated to waste transport long after changes in the resource picture call for an upgrading of the stream classification. While such a situation can be overcome by legislation, a loss of flexibility does result from the inertia inherent in a stream classification approach to water quality management.

A second difficulty, and one confronting all quality control systems, is that the minimum requirements cannot keep pace with the variety and change of industrial wastes. Certain quality factors, especially those typical of domestic wastes, are subject to precise numerical specification. Others, such as the as-yet-uninvented chemical, must be covered by broad cate-

Table 12-2 *Stream standards based on stream classification and waste water treatment*

Class	Use	Standards of quality at low-water stage	Required treatment of sewage	Emergency treatment
D (Bad)	For rough industrial uses and for irrigation	Absence of nuisance, odors and unsightly suspended or floating matters; dissolved oxygen present	Sedimentation, except in large receiving waters	Chlorination; ferric chloride treatment to remove hydrogen sulphide; addition of nitrate to supply oxygen
C	For fishing	D.O. content not less than 3 and preferably 5 ppm; CO_2 not more than 40 and preferably 20 ppm*	Sedimentation; chemical or biological treatment where necessary	Aeration; addition of diluting waters
B	For bathing, recreation, and shellfish culture	No visible sewage matters; a bacterial standard such as coliform density less than 100 per 100 ml	As in class C; chlorination if necessary	Chlorination
A (Good)	For drinking water after chlorination	In the absence of filtration,† a bacterial standard such as coliform density less than 50 per 100 ml. Chemical standards for substances not removable by common treatment methods	As in class B; removal of certain taste-producing substances such as phenols	Treatment of drinking water with heavy doses of chlorine and activated carbon

*At high temperatures the tolerance of fish to low D.O. and high CO_2 is decreased; high temperatures are also objectionable in themselves.

†With complete purification in modern filtration works, a bacterial standard such as 5,000 coliform density per 100 ml. will normally permit production of a safe drinking water.

SOURCE: Imhoff and Fair [4]. Reproduced by permission of the publisher.

Table 12-3 *Receiving water quality criteria based on use and stream classification*

Class	Use	
A	Water supply with chlorination only (fish life, bathing, recreation, process also)	Coliform bacteria MPN, 50/100 ml maximum Color < 10 ppm Turbidity < 10 ppm pH—6.0 to 8.0 range D.O. not less than 7.5 ppm; no single observation < 6.0 ppm No toxic substances, free acid, debris, odor or taste producers (except natural sources); no sludge deposits of any kind
B	Bathing, fish life, recreation; water supply after complete treatment	Coliform bacteria—MPN (monthly average) between 50 and 500/100 ml Color—20 ppm maximum desirable Turbidity—40 ppm maximum desirable pH—6.0 to 8.5 5-day BOD—maximum any sample = 3.0 ppm, average < 1.5 ppm D.O.—monthly average > 6.5 ppm; no sample < 5.0 ppm Other conditions—same as class A
C	Water supply after complete treatment; industrial process, navigation, etc.	Coliform bacteria—monthly average, 500 to 5000/100 ml Color and turbidity—removable by filtration pH—6.0 to 8.5 5-day BOD—monthly average < 2.0 ppm; single sample < 4.0 ppm D.O.—monthly average > 6.5; single sample > 5.0 Other conditions—same as class A
D	Navigation; cooling water, etc.	Shall not constitute nuisance pH—6.0 to 8.5 5-day BOD—monthly average < 3.0 ppm; single sample < 5.0 ppm Other conditions—no toxic substance, free acid, floating debris

gorical specifications, e.g., "no toxic substances." The result is that requirements always lag behind the fact of some quality factors, and quality management involves a continuous program of discovery of violation and proliferation of requirements.

CRITERIA BASED
ON BENEFICIAL USE

Progressive refinement of receiving water criteria from dilution to stream classification resulted from each new law, taking advantage of the obvious limitations of its predecessors as the necessity for water quality control moved west from the England of 1900 to the United States of 1950. Consequently, when the Pacific Coast was confronted with the need for pollution control legislation, a considerable body of experience was available to legislators.

In such a circumstance, California has deviated appreciably from the more customary type of standards of receiving water quality which depend on the beneficial use to which the water is to be put, in that it does not sanction the grouping of receiving waters into classes A, B, C, D, etc. While its rulings may in time have a similar practical effect, the classification of water quality will be controlled by the land and water use growth rather than by some preconceived, and often arbitrary, system of classification. Hence the quality requirements imposed are subject to continuously increasing restrictions as conditions warrant.

California's water pollution control act of 1949 [5] has two distinct goals:

1. Protection of the quality of the waters of the state for present and future beneficial use
2. Maximum use of those waters for waste disposal

The act set up a State Water Quality Control Board of 14 men, divided the state into 9 regions, and provided for a 5-man regional board in each region. The objective of the state board is

the prevention and control of pollution and contamination of the waters of the State at a minimum of expense consistent with obtaining this objective. In achieving this objective, it will be the policy of this board that its actions and those of the regional water pollution control boards shall be so directed as to secure that degree of care in the planning and operation of works for the treatment and disposal of sewage and industrial wastes as will adequately protect the public health and all of the beneficial uses of waters in this State and at the same time permit the legitimate planned usage of those waters for receiving suitably prepared wastes so that an orderly growth and expansion of cities and industries may be possible.

For practical reasons the Board does not displace the State Department of Public Health, a difficulty resolved by making a distinction between "contamination" (an immediate threat to health) and "pollution" (a factor inimical to other beneficial uses).

The State Board serves in a coordinating and appellate capacity to:

1. Formulate a statewide policy for control of water pollution with due regard to the authority of the regional boards
2. Review acts of a regional board where the regional board has failed to take or obtain appropriate action to correct a condition of pollution
3. Administer statewide programs of research or of financial assistance for water pollution control
4. Allocate funds for the administrative expenses of the regional boards

Each regional board is autonomous in its region. Each consists of representatives from the following: Water supply, irrigated agriculture, county government, city government, and industrial waste. As a regulatory agency the board has three principal duties:

1. Formulating and adopting long-range plans and policies
2. Setting and enforcing waste discharge regulations
3. Coordinating the interests of other agencies

In controlling pollution the board follows four steps:

1. Enunciates beneficial water uses which it intends to protect
2. Defines water quality criteria to protect beneficial water uses
3. Prescribes waste discharge requirements
4. Checks compliance and enforces requirements

In "enumerating beneficial water uses" the board takes cognizance of the fact that the law defines pollution as an impairment of the quality of water by sewage and industrial waste to a degree which *adversely* and *unreasonably* affects waters for domestic, industrial, agricultural, navigational, recreational, or other beneficial use.

The law establishes what constitutes beneficial water use—the board states only which of these water uses it intends to protect in any case. In doing this it (*a*) considers established water use, planned future uses, and need for economical waste disposal; (*b*) holds public hearings; and (*c*) takes action at a public meeting.

After setting forth the water use to be protected, the board must establish limits of chemical, physical, and biological characteristics beyond which impairment for beneficial use occurs. In setting these standards the board consults authorities in the proper fields (agriculture, water supply, recreation, etc.). It expects the criteria, with a reasonable margin of safety, to represent maximum concentrations of pollutants which may be achieved. A continuing study of water for indications of unreasonable impairment will exist and careful checks on waste discharges and enforcement procedures may be in order.

In carrying out the last two steps in the control of pollution—prescribing discharge requirements and enforcing these requirements—the board makes use of periodic checks and, if necessary, of the courts of law. Waste dischargers may appeal to the state board, which may hear or refuse to hear the case as its merits seem to warrant.

A greatly simplified hypothetical case is shown graphically in Figure 12-1. Here the beneficial uses and appropriate criteria are specified, as are the

Figure 12-1   *Concept of water quality control for beneficial use. (After Bacon and Sweet [6]; reproduced by permission of the American Society of Civil Engineers.)*

Beneficial use to be preserved in river water	Water quality criteria prescribed by control board		Possible conditions if pollution not controlled
Irrigation	700 ppm dissolved minerals, maximum 40 percent sodium ratio, maximum 0.5 ppm boron, maximum		High concentrations; crop damage and loss of production
Bathing	10 coliform bacteria per ml, maximum		High coliform counts; an actual hazard to public health (Contamination)
Fishing	5 ppm of dissolved oxygen (D.O.), minimum		Low dissolved oxygen; reduced fisheries resources or fish kills

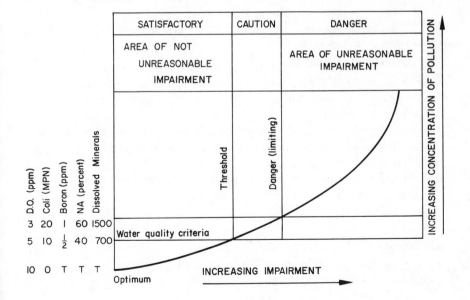

limits which define "not unreasonable" and "unreasonable" impairment of quality.

EVALUATION OF STANDARDS

In recent years the tendency has been to set standards for receiving waters rather than to specify the exact degree of treatment required for each municipal or industrial waste discharger. It is generally understood that the criteria shall serve as a guide in analyzing each problem of effluent discharge as it arises. Furthermore, they are intended to apply only for some reasonable time in the future. Thereafter, a changed public policy may require upgrading the requirements for any stream—a situation not easily accomplished under stream classification laws.

In establishing criteria or standards, consideration should include the following:

1. Not all streams were of the same quality before pollution began.
2. All beneficial uses of the stream, including waste disposal, should be taken into account.
3. In a dynamic civilization, standards must be revised from time to time to meet changed needs.

Serious disadvantages of any stream standard include the facts that:

1. Minimum permissible quality of water tends to become its maximum quality, as waste dischargers seek to get by with the least possible treatment that will meet the standards.
2. Standards tend to become fixed values rather than guides to judgment, hence may result in arbitrary decisions.
3. If stream classification is involved, future use of the stream tends to be frozen in advance.

The advantages include such considerations as:

1. The industry or town fortunately located on a stream or reach of water having a large assimilative capacity is not penalized, as is the case with effluent standards.
2. The degree of treatment imposed on a small industry with a small waste volume is less than would be required for a large installation of the same type in the same location.
3. Provision may be made for an orderly revision of regulations as conditions require, with less administrative difficulty.

REFERENCES

1   J. E. McKee and H. W. Wolf: "Water Quality Criteria," 2d ed., Report to California State Water Quality Control Board SWPCB Publication 3A, 1963.

2   Frank M. Stead: *Proceedings, Joint Seminar on Water Quality Research,* California State Department of Public Health, Bureau of Sanitary Engineering, June, 1965.

3   J. E. McKee: "Water Quality Criteria," California State Water Quality Control Board Publication 3, 1952.

4   Karl Imhoff and Gordon Fair: *Sewage Treatment,* 1st ed., John Wiley & Sons, Inc., New York, 1940.

5   *Report of the Interim Fact-Finding Committee on Water Pollution,* State of California, 1949.

6   V. W. Bacon and C. A. Sweet: "Setting Water Quality Criteria in California," ASCE Separate No. 698, May, 1955.

CHAPTER 13

Quality requirements
of beneficial uses

INTRODUCTION

The minimum objective of the quality control procedures discussed in Chapter 12 is to maintain the quality of the resource pool at such a level that it can be utilized directly in agriculture, recreation, etc., or upgraded for industrial or domestic use by current technology of water purification. Thus both the objectives and the engineering of water quality management depend upon a knowledge of what factors, and what concentrations of each, can be tolerated by the various beneficial uses which management seeks to protect.

QUALITY REQUIREMENTS
OF DOMESTIC USE

The quality of water which must pertain in the resource pool if it is to be suitable for domestic use varies widely with the individual quality factor under consideration. In general, it may be said that as a matter of principle:

1. To be suitable for domestic purposes the raw water supply must be of such quality that the Public Health Service Drinking Water Standards [1], in all factors where absolute limits are specified, are achievable by normal water treatment technology, i.e., sedimentation, coagulation, filtration, chlorination.

2. The quality standards which are recommended but not obligatory should be exceeded only as a result of characteristics of the natural water and not as a result of the increments added by return waste waters; for example, softening or deionization, or simply human tolerances of such things as salinity and hardness, should not be imposed as a result of quality depression by previous beneficial use of the raw water.

The Public Health Service Drinking Water Standards, originally applicable to common carriers in interstate traffic, have long since become

132

the basis of all state standards for domestic water supply, although in a few cases the state requirements may be more restrictive. They have been derived from all the various bases noted in Chapter 12, from attainability to biostatistics. The standards present a considerable amount of explanatory data on general environmental sanitation, sampling techniques and frequency, and methods of analysis. For coliform bacteria and certain chemical factors, maximum permissible limits are specified. Concentrations in excess of these limits are cause for action by the health department in rejecting the water for human consumption. Physical characteristics and several chemical factors are subject to recommended limits which are in no sense absolute, and in many cases less than that naturally occurring in the local water supply. Adherence to these limits would permit people freely to move about the country without danger of digestional upsets or aesthetic objection due to the local water supply. A full presentation of the significance of various quality factors is beyond the scope of this work but may be found in the monumental work of McKee and Wolf [2].

Standards of physical quality of water call for the absence of materials offensive to the senses of sight, taste, and smell. To this end, limits are recommended for each of certain physical factors in terms of numerical values on an arbitrary scale:

Turbidity	5 units
Color	15 units
Threshold odor number	3 units

In the case of turbidity, the recommended value is somewhat less than the probable aesthetic limit because of the difficulty of obtaining good chlorine penetration of particles which may cause turbidity.

Biological standards, although founded on statistical probability, represent maximum permissible rather than recommended maximum values. The standards, which involve considerable detail, are summarized in Table 13-1.

Table 13-1 *Summary of bacteriological standards of water quality*

Sample examined	Limits*
Standard 10-ml portions	Not more than 10 percent in one month shall show coliforms. (Subject to further specified restrictions.)
Standard 100-ml portions	Not more than 60 percent in one month shall show coliforms. (Subject to further specified restrictions.)

*Public Health Service Drinking Water Standards [1].

In general, the values specified mean that the most probable number (MPN) of coliform organisms shall not exceed 1 coliform organism per 100 ml of water. ~ simply indicator organism

Chemical standards, both recommended and maximum permissible, are summarized in Table 13-2. For the purpose of comparison, both the 1946 and the current standards as revised in 1961 are included in the table. In some instances the comparison bespeaks the immutability of a standard, or perhaps the slow progress in establishing epidemiological bases for standards. Others such as selenium and lead were drastically reduced on the basis of evidence accrued in the 1946-1961 period. In the case of cyanide and arsenic the 1961 recommended maximum is appreciably less than the maximum permissible value. Chemicals appearing on the list for the first time in 1961 include ABS, silver, nitrate, iron and manganese as separate materials, and carbon chloroform extract. This indicates both a lengthening of the spectrum of quality factors considered significant in domestic water supply, and an emerging question of the importance of a vast number of organic compounds, such as pesticides, presently lumped in the single category of "exotic organic chemicals."

Standards of radioactivity require that added radiation shall not bring the total from all sources above the maxima specified by the Federal Radiation Council and approved by the President. Water supplies shall be approved without consideration of other sources of radiation from radium-226 and strontium-90 when the water contains these substances in amounts not exceeding 3 and 10 $\mu\mu$Ci/liter, respectively.

In the known absence of strontium-90 and alpha emitters, the water supply is acceptable when the gross beta concentrations do not exceed 1,000 $\mu\mu$Ci/liter. In excess of this amount the water shall be rejected for public use.

Although a detailed evaluation of the limitations of water-treatment processes is reserved to Chapter 20, it is important to note here that only the physical characteristics, the bacterial count, and perhaps the TDS of water are changed by the normal processes of water purification. Control of all the other factors listed in the Public Health Service Drinking Water Standards must be achieved by a combination of dilution and treatment of return flows from beneficial uses. That is to say, the quality requirements of domestic use must to a large extent be met by management of the quality of the water resource pool.

QUALITY REQUIREMENTS
OF INDUSTRY

By far the most varied spectrum of quality requirements of beneficial uses is found in industrial water needs. Table 13-3 gives an idea of the water quality tolerances for several industrial applications. A study of the table leads to the conclusion that the needs of industry cannot possibly become the criterion by which the quality of the freshwater resource is managed. Certainly a water with the 0 oxygen needed in high-pressure boiler feed, the pH limit of 6.5-7 of brewing, and the 0.3 turbidity limit of rayon manufacture would be a weird concoction indeed, unsuited to most uses, and difficult to handle.

Table 13-2 *Chemical standards of drinking water*

Quality factor	Recommended maximum limits,* mg/liter		Maximum permissible† concentrations, mg/liter	
	1946	1961 Revision	1946	1961 Revision
Alkylbenzenesulfonate (detergent)	...	0.5		
Arsenic	...	0.01	0.05	0.05
Barium			...	1.0
Cadmium	...	...	...	0.01
Carbon chloroform extract (exotic organic chemicals)	...	0.2		
Chloride	250	250		
Chromium	...	...	0.05	0.05
Copper	3.0	1.0		
Cyanide	...	0.01	...	0.02
Fluoride	...	‡1.7	1.5	‡2.2
Iron plus manganese	0.3			
Iron	...	0.3		
Lead	...	...	0.1	0.05
Manganese	...	0.05		
Nitrate	...	45		
Phenols	0.001	0.001		
Selenium	...	...	0.15	0.01
Silver	...	...	...	0.05
Sulfate	250	250		
Total dissolved solids (TDS)	500	500		
Zinc	15	5		

*Concentrations in water should not be in excess of these limits, when more suitable supplies can be made available.

†"Maximum permissible" implies that which constitutes grounds for rejection of supply.

‡Fluoride temperature concentration relationships are discussed in detail in the text.

Table 13-3 *Typical quality factor limits for selected industries (Values in mg/liter)*

(handwritten note: Shun or bd. / Values not important)

Type of industry	Turbidity	Color	Hardness	Alkalinity	pH	Total dissolved solids	Fe and Mn
Baking	10	10	...	...	...	...	0.5
Boiler feed*							
0–150 psi	20	80	75	...	8.0+	3000–1000	
150–250 psi	10	40	40	...	9.5+	2500–500	
>250 psi	5	5	8	...	9.0+	1500–100	
Brewing†							
Light	10	...	...	75	6.5–7	500	0.1
Dark	10	...	...	150	7.0→	1000	0.1
Canning							
General	10	...	...	...	...	...	0.2
Legumes	10	...	25–75	50	...	...	0.2
Carbonated beverages	2	10	250	...	...	850	0.3
Confectionery	...	...	50	...	...	100	0.2
Cooling	50	...	50	...	...	...	0.5

	1–5	5		30–50		300	
Ice making						300	0.2
Laundering	...		50			200	0.2
Clear plastics	2	2	...				0.02
Pulp and paper							
Groundwood	50	20	180			300	1.0
Kraft	25	15	100			200	0.2
Soda and sulfite	15	10	100			200	0.1
Light paper	5	5	50				0.1
Rayon (viscose)							
Production	5	5	8	50			0.05
Manufacture	0.3	...	55	...	7.8–8.3	...	0.0
Tanning	20	10–100	50–135	133	8.0	...	0.2
Textiles							
General	5	20	20				0.25
Dyeing	5	5–20	20				
Wool scouring		70	20				1.0
Cotton bandage	5	5	20				0.2

*No D.O.; Al_2O_3, 5–0.5; SiO_2, 40–5; CO_2, 200–40; HCO_3, 50–5; OH, 50–30.

†Ca, 100–500; $CaSO_4$, 100–500; low odor.

SOURCE: American Water Works Association [3]. Reproduced by permission.

Therefore it would seem axiomatic that industry will have to continue to treat water in accord with its own needs for process water, and to locate its plants where the local resource can be suited to its process and cooling water requirements. Generally the siting of a new plant is governed more by the quality of return water that industry is permitted to discharge than by the quality of the water at its intakes. However, the movement of industrial plants from the heavily industrialized Northeast, for example, has often been triggered by the low quality of the intake water as a result of upstream discharge of wastes by numerous other industrial establishments.

QUALITY REQUIREMENTS
OF AGRICULTURE

In a previous chapter attention was called to the fate of past civilizations which have depended upon the irrigation of crops for their prosperity, and it was speculated that if ours should go the same route it may well be because intensive multiple repeated use of the water resource will have salted the water rather than the soil. In any event, it is of more than normal concern to us that the quality of the freshwater pool be suited to the needs of agriculture.

The factors that limit the usefulness of a water for agriculture, and the concentrations at which their effects are felt either in a minimal or a catastrophic way, have been the object of much research and many publications. Only a few of the more important factors which define the quality of water acceptable to growing of crops are outlined herein. They are, of course, based on the limitations of current species of plants and presuppose the maintenance of a situation in which the choice of crops to be planted by the farmer is governed by factors other than water quality. At the present state of our national development, one of the goals of water quality management should be the maintenance of such a freedom as long as possible.

Salinity   *main consideration*

Eldridge, in his summary of "quality considerations" [4], states that "Criteria for irrigation water are developed on the basis of salinity, sodium, boron, and bicarbonate concentrations.... Irrigation water should be relatively high in calcium and magnesium...; boron in irrigation water should not exceed one to two ppm...; and silica, nitrate, and fluoride are undesirable in domestic water but in the normal concentrations present no hazard to irrigation water."

Salinity is measured in terms of electrical conductivity (EC_e) expressed in millimhos per centimeter (mmhos/cm) at 25°C. (The average relationship between EC_e and salt concentration is: 1 mmho/cm $\cong$ 640 mg/liter salt.)

1000 → 1500 mg/l chlorides
will not effect crops
except salt greengrass

Observation of EC_e is made on a solution called the "saturation extract" withdrawn from a soil sample which has been saturated with distilled water. Since plants differ in their tolerance to salinity, crops may be classified in broad categories in accord with their response to salinity. Data from the U.S. Department of Agriculture [5] showing the relationship of EC_e to plant yields are shown in Table 13-4.

Table 13-4 *Crop response to salinity*

Salinity (EC_e, mmhos/cm at 25°C)	Crop response
0–2	Salinity effects mostly negligible.
2–4	Yields of very sensitive crops may be restricted.
4–8	Yields of many crops restricted.
8–16	Only tolerant crops yield satisfactorily.
Above 16	Only a few very tolerant crops yield satisfactorily.

SOURCE: U.S. Department of Agriculture [5]. Reproduced by permission.

Eldridge [2] notes that "about one-half of the waters used for irrigation in the west comes within the range 0.25 to 0.75 mmho/cm. ...." In terms of dissolved minerals this is about 160 to 480 mg/liter.

The Department of Agriculture reports [5] that both growth and yield of plants are progressively decreased as salinity is increased above the threshold of their salt sensitivity. Growth reduction is characterized by smaller than normal leaves, stems, and fruits. However, with some plants such as barley, wheat, and cotton in which the desired product is the seed or fiber pod, a reduction in plant size as much as 50 percent may not significantly reduce the yield. Both an interference with plant nutrition and acute injury to the plant may be involved in the salinity effect.

Salt tolerances of field, vegetable, and forage crops have been graphically summarized by the Department of Agriculture [5] as shown in Figure 13-1, which indicates also the salinity at which 10, 25, and 50 percent reductions in yield might be anticipated.

Eldridge notes that "plants have difficulty in obtaining water from saline solutions. The characteristic(s) of soils, however, are not adversely affected by high concentrations of salts, if sodium is low in comparison with calcium and magnesium. Sodium renders soils impermeable to air and water; and when wet, these soils become plastic and sticky. The effect of sodium on the soil is measured by the 'sodium-absorption ratio' (SAR) which is the

Figure 13-1 *Salt tolerance of crops.
(Reproduced from Agriculture
Information Bulletin No. 283 [5] by
courtesy of the U. S. Department
of Agriculture.)*

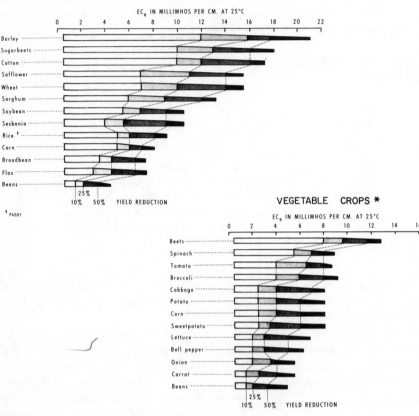

ratio of sodium ion to calcium and magnesium ions." The formula used is

$$SAR = \frac{Na}{\sqrt{\dfrac{Ca + Mg}{2}}}$$

A value of SAR = 8 is considered satisfactory, 12 to 15 is marginal, and more than 20 is serious and requires special management practices, e.g., applying gypsum ($CaSO_4$) to the soil or treating the soil with acid to release calcium from insoluble lime.

"SAR must be used in combination with total salt content since the higher the salts the more sodium can be tolerated." This fact has led to a classification of irrigation waters on a scale of C_1 to C_4, as follows:

1. Low salinity water (C_1) can be used on most crops and soils.
2. Medium salinity water (C_2) can be used on most crops but requires a moderate amount of leaching.
3. High salinity water (C_3) requires good drainage, permeable soils, and salt-resistant crops.
4. Very high salinity water (C_4) is not normally good for irrigation.

Boron

Boron is toxic to plants in certain concentrations but varies with species and climate factors. In general, the concentrations are distributed as follows:

Type of crop	mg/liter
Sensitive (fruits, nuts, beans)	0.33–1.25
Semitolerant (cereals, vegetables, cotton)	0.67–2.50
Tolerant (alfalfa, sugar beets, asparagus)	1.00–3.75

Bicarbonates and carbonates

Eldridge [4] notes that:

Bicarbonates in irrigation water tend to render calcium more soluble. When calcium bicarbonate enters the soil an increase in temperature or evaporation may precipitate the calcium as $CaCO_3$ which tends to hold the calcium in the soil. This is of importance since it keeps the calcium content of the soil high. Thus, reduction of calcium in the drainage water results in an increase in the sodium-absorption ratio.

Some waters contain 'residual sodium carbonate' which is defined as the sum of the equivalents of normal carbonate and bicarbonate minus the sum of the equivalents of calcium and magnesium.

It is generally concluded that water containing less than 66 mg/liter residual normal sodium carbonate (Na_2CO_3) can be safely used in irrigation. Between 66 and 132 mg/liter is marginal; above 132 mg/liter is not suitable for agriculture.

From an extensive review of the literature, McKee and Wolf [6] suggest tentative criteria for judging the suitability of water for irrigation use. (See

Table 13-5.) They call especial attention to the fact that sampling must reflect any important variability in the concentration of the quality factors considered important to agricultural use.

TABLE 13-5 *Tentative guides for* ᵒⁿ   bd
evaluating the quality of water used for irrigation

Quality factor	Threshold concentration*	Limiting concentration†
Coliform organisms, MPN per 100 ml	1000‡	§
Total dissolved solids (TDS), mg/liter	500‡	1500‡
Electrical conductivity, μmhos/cm	750‡	2250‡
Range of pH	7.0-8.5	6.0-9.0
Sodium adsorption ratio (SAR)	6.0‡	15
Residual sodium carbonate (RSC), meq	1.25‡	2.5
Arsenic, mg/liter	1.0	5.0
Boron, mg/liter	0.5‡	2.0
Chloride, mg/liter	100‡	350
Sulfate, mg/liter	200‡	1000
Copper, mg/liter	0.1‡	1.0

*Threshold values at which irrigator might become concerned about water quality and might consider using additional water for leaching. Below these values, water should be satisfactory for almost all crops and almost any arable soil.

†Limiting values at which the yield of high-value crops might be reduced drastically, or at which an irrigator might be forced to less valuable crops.

‡Values not to be exceeded more than 20 percent of any 20 consecutive samples, nor in any 3 consecutive samples. The frequency of sampling should be specified.

§Aside from fruits and vegetables which are likely to be eaten raw, no limits can be specified. For such crops, the threshold concentration would be limiting.

SOURCE: McKee and Wolf [6]. Reproduced by permission of the California State Water Quality Control Board.

QUALITY REQUIREMENTS OF RECREATIONAL USE

Recreational use of water as herein discussed refers primarily to bathing, swimming, and other water-contact sports; boating; and aesthetic enjoyment. Fishing is, of course, a recreational use of water, but the quality considerations in that case are those required for the protection of aquatic life and are hence presented in that context.

The general criterion of quality for recreational use of water is obvious: freedom from obnoxious suspended or floating material, objectionable color, or foul odors. Furthermore, the water should be free of substances (including pathogens) which are dangerous to swallow.

Specific standards in terms of numbers, however, are not so easily established. Largely because of the need for some sort of a standard, coliform organisms have often been considered. The possibility of infection certainly exists when fecal organisms abound in the water, although British studies of the North Sea failed to show any correlation between disease and sewage-polluted water and beaches. The result is that standards set by regulatory agencies have generally been related to aesthetic considerations and feasibility of compliance rather than to any sound epidemiological evidence. A general disinclination of people to come in contact with sewage, together with uncertainty of the mode of transmission of infectious hepatitis, suggests that some coliform standard or criterion should be utilized. Natural agents such as the organisms which produce schistosome dermatitis and leptospirosis occur in some waters and so render them unfit for recreation involving contact, but the methods of preventing such infestation of waters do not appear in standards.

Table 13-6, compiled from data presented by McKee and Wolf [6], reflects the rather uncertain quality limits suggested by various agencies for recreational waters.

TABLE 13-6 *Tentative guides for evaluating recreational waters*

Determination	Water contact		Boating and aesthetic	
	Noticeable threshold	Limiting threshold	Noticeable threshold	Limiting threshold
Coliforms, MPN per 100 ml	1000*	†		
Visible solids of sewage origin	None	None	None	None
ABS (detergent), mg/liter	1*	2	1*	5
Suspended solids, mg/liter	20*	100	20*	100
Flotable oil and grease, mg/liter	0	5	0	10
Emulsified oil and grease, mg/liter	10*	20	20*	50
Turbidity, silica scale units	10*	50	20*	‡
Color, standard cobalt scale units	15*	100	15*	100
Threshold odor number	32*	256	32*	256
Range of pH	6.5-9.0	6.0-10.0	6.5-9.0	6.0-10.0
Temperature, maximum °C	30	50	30	50
Transparency, Secchi disk, ft	...	...	20*	‡

*Value not to be exceeded in more than 20 percent of 20 consecutive samples, nor in any 3 consecutive samples.

†No limiting concentration can be specified on the basis of epidemiological evidence, provided no fecal pollution is evident. (Note: Noticeable threshold represents the level at which people begin to notice and perhaps to complain. Limiting threshold is the level at which recreational use of water is prohibited or seriously impaired.

‡No concentrations likely to be found in surface waters would impede use.

SOURCE: McKee and Wolf [6]. Reproduced by permission of the California State Water Quality Control Board.

QUALITY REQUIREMENTS
FOR LIVESTOCK AND WILDFOWL

Animals react very much the same as humans to highly mineralized waters. Generally no animal will choose to drink saline water when better water is available. Sudden changes from slightly mineralized to highly mineralized waters have been known to kill livestock from salt poisoning. However, the tolerance of animals depends upon many factors including health, salt content of the diet, and the nature of the salts involved. Indications are that the maximum concentrations of salts that can be tolerated by domestic animals without danger of injury by osmotic effect lie between 15,000 and 17,000 mg/liter. However, a value of 10,000 mg/liter is more realistic for sheep and perhaps 7,000 mg/liter for milk cows in production. Obviously, these values are considerably higher than can be tolerated by humans.

In recent years high nitrate levels in water have been considered responsible for losses in milk and meat production, for stock-breeding difficulties, and for the death of many farm animals, particularly ruminants. Fincher [7] reported that dairy cows died from drinking water containing 7,000 mg/liter of sodium nitrate. Winks et al. [8] reported the death of pigs fed with soup prepared with groundwater containing 1,740 to 2,970 mg/liter of this same chemical. Lethal doses of sodium nitrite, derived from the nitrate by the animals' digestive processes, were found to be 0.09 g/kg of body weight of pigs. However, because plants may also supply enough nitrogen to poison animals [9], nitrate-free water is recommended. Some investigators believe that 50 to 100 mg/liter is too much nitrate. Values in the 200- to 400-mg/liter range are suggested in Table 13-7 on the basis of extensive literature.

Toxic blue-green algae cultured by sewage in water have been the cause of death by poisoning of livestock, poultry, and wild birds and animals. Gorham [11] cites an instance of cattle dying within 1.5 hours after drinking from a lake containing a concentrated water bloom of blue-green algae. Wastes from dairies and slaughterhouses have been suspect in the spreading of animal diseases through water. Oily substances are detrimental to livestock and the unavailability of freshwater forces them to drink it. Wildfowl losses are particularly great when contact with oil fouls up their feathers.

Wildfowl poisoning has been reported to result from pesticides concentrated in the food chain, and from highly polluted water. Protection of wildfowl, therefore, is best attained by maintaining a water quality suited to aquatic life.

QUALITY REQUIREMENTS
OF AQUATIC LIFE

The extensive literature on the effects of pollution on aquatic life is such as almost to discourage one who would summarize it. As has been previously

TABLE 13-7 *Tentative guides for evaluating the quality of water used by livestock*

Quality factor	Threshold concentration*	Limiting concentration†
Total dissolved solids (TDS), mg/liter	2500	5000
Cadmium, mg/liter	5	
Calcium, mg/liter	500	1000
Magnesium, mg/liter	250	500‡
Sodium, mg/liter	1000	2000‡
Arsenic, mg/liter	1	
Bicarbonate, mg/liter	500	500
Chloride, mg/liter	1500	3000
Fluoride, mg/liter	1	6
Nitrate, mg/liter	200	400
Nitrite, mg/liter	None	None
Sulfate, mg/liter	500	1000‡
Range of pH	6.0-8.5	5.6-9.0

*Threshold values represent concentrations at which poultry or sensitive animals might show slight effects from prolonged use of such water. Lower concentrations are of little or no concern.

†Limiting concentrations based on interim criteria, South Africa. Animals in lactation or production might show definite adverse reactions.

‡Total magnesium compounds plus sodium sulfate should not exceed 50 percent of the total dissolved solids.

SOURCE: McKee and Wolf [6]. Reproduced by permission of the California State Water Quality Control Board.

pointed out, anything that interrupts the food chain at any level is serious to all organisms higher up the chain. Thus a lack of fertilizers may lead to an aquatic desert. However, as discussed in Chapter 11, the likelihood is quite remote that man's activities are going to wipe out the fertilizers in water where they already exist. The danger lies in toxicity acting in different ways on different creatures, or in suffocation of organisms as a result of lowered or completely obliterated oxygen content of the water. It is for this reason that BOD and oxygen relationships have been featured most prominently in assessing the quality of water for aquatic life, and why domestic sewage has been the first line of attack whenever the protection of aquatic life is one of the beneficial uses of water to be protected by public agencies. Quality management for aquatic life therefore revolves about oxygen; although with the advent of organic pesticides, their effects on the water resource have been a matter of concern.

Bacterial pollution of water has been a factor in relation to aquatic life, not because of its effects on the organisms themselves but rather because of the danger to human beings from eating raw shellfish or by contaminating the food or drink of the fisherman through carelessness. Shellfish, being nature's little sewage treatment plants, can concentrate bacteria and evidently virus to a degree that is dangerous to the gourmet.

On the other hand, the possibility that wastes may produce disease in fishes cannot be ruled out. Recently it has been found that the English sole in San Francisco Bay caught near the sewer outfalls have growths on their heads and bodies. Virus in wastes is suspect but the results of investigative work are not yet in. Nevertheless it seems evident that some quality factor in the Bay water is the cause.

McKee and Wolf, in Table 13-8, summarize tentative guides for evaluating water quality in the context of fish propagation. They also summarize work which leads to the general conclusion that salinities below 12,000 mg/liter are definitely detrimental to oysters; that oysters and clams are critically affected by temperature but different species have different ranges of tolerance. Concentration of radionuclides and, in some cases, of exotic organics, is also a factor in considering the suitability of water for aquatic life.

Klein [10] lists a series of coal tar derivatives which lead to objectionable tainting of fish flesh, and which at the same time are toxic to fish. He notes that phenols, chlorphenols, and similar compounds from industrial wastes may thus damage the value of fisheries. He also presents a quite comprehensive list of some 200 reports on the effect of various substances on fish. A recent publication (1965) of the Robert A. Taft Center [11] gives a wealth of information on the response of lower creatures in the aquatic food chain—particularly the aquatic insects, protozoa, and bacteria, as indicated by numbers of organisms present—to a whole spectrum of ions, compounds, and environmental conditions. These findings, and those of the numerous references from which they are drawn, are too voluminous to summarize here and to interpret as quantitative measures of quality. The important factor is that a great deal of investigative work is being directed to the matter of the water quality needs of aquatic organisms, and it may be expected that as time goes on and the data become more conclusive, such quality considerations will be reflected in an increasing stringency of requirements to be met by beneficial users of the water resource which return used water to the resource pool.

For the present purpose, Table 13-8, together with Table 13-9 derived from a number of sources [10, 11, 12, 13, 14], may serve to indicate the general criteria by which water can be judged as to its suitability for aquatic life.

The data presented in Table 13-9 are but a small fraction of that available in the literature. They are, of course, not strictly comparable because both the organisms tested and the test conditions are highly variable. Some repre-

TABLE 13-8 *Tentative guides for evaluating the quality of water*

Determination	Threshold concentration*	
	Freshwater	Saltwater
Total dissolved solids (TDS), mg/liter	2000†	
Electrical conductivity, μmhos/cm @ 25°C	3000†	
Temperature, maximum °C	34	34
Maximum for salmonoid fish	23	23
Range of pH	6.5-8.5	6.5-9.0
Dissolved oxygen (D.O.), minimum mg/liter	5.0‡	5.0‡
Flotable oil and grease, mg/liter	0	0†
Emulsified oil and grease, mg/liter	10†	10†
Detergent, ABS, mg/liter	2.0	2.0
Ammonia (free), mg/liter	0.5†	
Arsenic, mg/liter	1.0†	1.0†
Barium, mg/liter	5.0†	
Cadmium, mg/liter	0.01†	
Carbon dioxide (free), mg/liter	1.0	
Chlorine (free), mg/liter	0.02	
Chromium, hexavalent, mg/liter	0.05†	0.05†
Copper, mg/liter	0.02†	0.02†
Cyanide, mg/liter	0.02†	0.02†
Fluoride, mg/liter	1.5†	1.5†
Lead, mg/liter	0.1†	0.1†
Mercury, mg/liter	0.01	0.01
Nickel, mg/liter	0.05†	
Phenolic compounds, as phenol, mg/liter	1.0	
Silver, mg/liter	0.01	0.01
Sulfide, dissolved, mg/liter	0.5†	0.5†
Zinc, mg/liter	0.1	

*Threshold concentration is value that normally might not be deleterious to fish life. Waters that do not exceed these values should be suitable habitats for mixed fauna and flora.

†Values not to be exceeded more than 20 percent of any 20 consecutive samples, nor in any 3 consecutive samples. Other values should never be exceeded. Frequency of sampling should be specified.

‡Dissolved oxygen concentrations should not fall below 5.0 mg/liter more than 20 percent of the time and never below 2.0 mg/liter. (Note: Recent data indicate also that rate of change of oxygen tension is an important factor, and that diurnal changes in D.O. may, in sewage-polluted water, render the value of 5.0 of questionable merit.)

SOURCE: McKee and Wolf [6]. Reproduced by permission of the California State Water Quality Control Board.

TABLE 13-9 *Observed lethal concentration of selected chemicals in aquatic environments*

Chemical	Organism tested	Lethal concentration, mg/liter	Exposure time, hr	Reference number
ABS (100 percent)	Fathead minnow	3.5-4.5	96	15
ABS (100 percent)	Bluegills	4.2-4.4	96	15
Household syndets	Fathead minnow	39-61	96	15
Alkyl sulfate	Fathead minnow	5.1-5.9	96	15
LAS (C12)	Bluegill fingerlings	3	96	16
LAS (C14)	Bluegill fingerlings	0.6	96	16
Acetic acid	Goldfish	423	20	17
Alum	Goldfish	100	12-96	17
Ammonia	Goldfish	2-2.5 NH_3	24-96	17
Ammonia	Perch, roach, rainbow trout	3 N	2-20	18
Sodium arsenite	Minnow	17.8 As	36	19
Sodium arsenate	Minnow	234 As	15	19
Barium chloride	Goldfish	5000	12-17	17
Barium chloride	Salmon	158	. . .	20
Cadmium chloride	Goldfish	0.017	9-18	21
Cadmium nitrate	Goldfish	0.3 Cd	190	22
CO₂	Various species	100-200	. . .	23
CO	Various species	1.5	1-10	23
Chloramine	Brown trout fry	0.06	. . .	24
Chlorine	Rainbow trout	0.03-0.08	. . .	25-26
Chromic acid	Goldfish	200	60-84	17

149

Compound	Species	Concentration	Time	Ref
Copper sulfate	Stickleback	0.03 Cu	160	10
Copper nitrate	Stickleback	0.02 Cu	192	27
Cyanogen chloride	Goldfish	1	6-48	28
H_2S	Goldfish	10	96	17
HCl	Stickleback	pH 4.8	240	22
HCl	Goldfish	pH 4.0	4-6	17
Lead nitrate	Minnow, stickleback, brown trout	0.33 Pb	. . .	29
Mercuric chloride	Stickleback	0.01 Hg	204	22
Nickel nitrate	Stickleback	1 Ni	156	22
Nitric acid	Minnow	pH 5.0	. . .	29
Oxygen	Rainbow trout	3 cc/liter	. . .	30
Phenol	Rainbow trout	6	3	31
Phenol	Perch	9	1	31
Potassium chromate	Rainbow trout	75	60	19
Potassium cyanide	Rainbow trout	0.13 Cn	2	31
Sodium cyanide	Stickleback	1.04 Cn	2	32
Silver nitrate	Stickleback	70 K	154	22
Sodium fluoride	Goldfish	1000	60-102	17
Sodium sulfide	Brown trout	15	. . .	33
Zinc sulfate	Stickleback	0.3 Zn	120	17
Zinc sulfate	Rainbow trout	0.5	64	34

Pesticides
1. Chlorinated hydrocarbons

Compound	Species	Concentration	Time	Ref
Aldrin	Goldfish	0.028	96	39
DDT	Goldfish	0.027	96	35
DDT	Rainbow trout	0.5-0.32	24-36	36-37
DDT	Salmon	0.08	36	37

TABLE 13-9 *Observed lethal concentration of selected chemicals in aquatic environments (continued)*

Chemical	Organism tested	Lethal concentration, mg/liter	Exposure time, hr	Reference number
Pesticides (*continued*)				
DDT	Brook trout	0.032	36	37
DDT	Minnow, guppy	0.75 ppb	29	38
DDT	Stoneflies (species)	0.32–1.8	96	11
BHC	Goldfish	2.3	96	39
BHC	Rainbow trout	3	96	40
Chlordane	Goldfish	0.082	96	39
Chlordane	Rainbow trout	0.5	24	36
Dieldrin	Goldfish	0.037	96	39
Dieldrin	Bluegill	0.008	96	39
Dieldrin	Rainbow trout	0.05	24	36
Endrin	Goldfish	0.0019	96	39
Endrin	Carp	0.14	48	41
Endrin	Fathead minnow	0.001	96	39
Endrin	Various species	0.03–0.05 ppb	. . .	42
Endrin	Stoneflies (species)	0.32–2.4 ppb	96	11
Heptachlor	Rainbow trout	0.25	24	36
Heptachlor	Goldfish	0.23	96	39
Heptachlor	Bluegill	0.019	96	39
Heptachlor	Redear sunfish	0.017	96	11
Methoxychlor	Rainbow trout	0.05	24	36
Methoxychlor	Goldfish	0.056	96	39

Pesticides (*continued*)

Toxaphene	Rainbow trout	0.05	24	36
Toxaphene	Goldfish	0.0056	96	39
Toxaphene	Carp	0.1	. . .	43
Toxaphene	Goldfish	0.2	24	44
Toxaphene	Goldfish	0.04	170	44
Toxaphene	Minnows	0.2	24	44
2. Organic phosphates				
Chlorothion	Fathead minnow	3.2	96	39
Dipterex	Fathead minnow	180	96	39
EPN	Fathead minnow	0.2	96	39
Guthion	Fathead minnow	0.093	96	39
Guthion	Bluegill	0.005	96	39
Malathion	Fathead minnow	12.5	96	39
Parathion	Fathead minnow	1.4-2.7	96	39
TEPP	Fathead minnow	1.7	96	39
3. Herbicides				
Weedex	Young roach and trench	40-80	1 month	11
Weeda Zol		15-30	1 month	11
Weeda Zol T.L.		20-40	1 month	11
Simazine (no plants present)	Minnow	0.5	< 3 days	11
Atrazine (A361) (plants present)	Minnow	5.0	24	11
Atrazine in Gesaprime	Minnow	3.75	24	11
4. Bactericides				
Algibiol	Minnow	20	24	11
Soricide tetraminol	Minnow	8	48	11

sent total kill of the test organism; others the TL_m, or 50 percent mortality time of a standard bioassay test. Temperature of water, age of test organism, and numerous other factors differ. Nevertheless, the table shows clearly that pesticide and toxic ions, the quality factors most difficult to measure and the least observed in the past, are the most critical to aquatic life. Hence the objective of preservation of aquatic life requires the development of a new set of parameters of water quality for management purposes.

Most of the concern herein expressed has been with the freshwater resource, whereas shellfish are brackish or saltwater organisms. It should be remembered that any program of quality management that affects the nature and rate of discharges from the freshwater resource pool into estuarial waters has a profound effect upon aquatic life. As noted in Chapter 7, water of any degree of purity constitutes a pollutant if introduced, either by design or inadvertence, into a saline environment. The effects may be either beneficial or adverse. For example, unusual amounts of floodwater in the early 1960s caused a very high kill of shellfish in Mobile Bay due to lowered salinity. Flood control may well reduce the hazards of life in estuarine waters. On the other hand, it may change the ecology of the marine resource beyond the shoreline. The effects on upstream management of the freshwater resource on the marine environment have only recently become the objective of oceanographers and marine biologists. Undoubtedly their findings will reveal further uncharted ramifications of nature's search for new equilibria as a result of man's efforts to manage the freshwater resource.

OTHER BENEFICIAL USES

The spectrum of beneficial uses of water extends beyond domestic, industrial, agricultural, recreational uses, and the preservation of useful or even useless aquatic life. It includes also navigation, power production, flood control, and wastes transportation. General statements might be made about the quality of water required for each of these benefits, but they would be relatively meaningless from a practical viewpoint. Specifically, the quality control measures necessary to protect other beneficial uses today would become overriding long before metal ships or turbine runners were affected.

SUMMARY

The general quality requirements for any sector of the water resource pool in relation to five levels of beneficial use are summarized in Table 13-10 on pages 156–161. Water of such quality is considered to be amenable to conventional treatment systems associated with beneficial use for domestic and industrial purposes, and acceptable to other beneficial uses which must forego pretreatment. Thus the table represents the minimum objectives which must be attained through treatment of return flows from beneficial uses by the engineered systems discussed in later chapters.

REFERENCES

1 *Public Health Service Drinking Water Standards*, U.S. Public Health Service, revised 1961.

2 J. E. McKee and H. W. Wolf: "Water Quality Criteria," 2d ed., Report to California State Water Quality Control Board, SWPCB Publication 3A, 1963.

3 *Water Quality and Treatment*, AWWA Manual, 2d ed., American Water Works Association, New York, 1950.

4 E. F. Eldridge: "Return Irrigation Water—Characteristics and Effects," U.S. Public Health Service, Region IX, May 1, 1960.

5 Leon Bernstein: "Salt Tolerance of Plants," Agriculture Information Bulletin 283, U.S. Department of Agriculture, 1964.

6 J. E. McKee and H. W. Wolf: "Water Quality Criteria," unpublished report to Paul R. Bonderson, California State Water Quality Control Board, 1963.

7 M. G. Fincher: *Cornell Vet.*, 26, 1936.

8 W. R. Winks, A. K. Sutherland, and R. M. Salisbury: "Nitrite Poisoning in Pigs," *Queensland. J. Agric. Sci.*, 71; *Water Pollution Abstr.*, 25(10), Abstr. No. 1528, 1950.

9 J. M. Tucker et al.: "Nitrate Poisoning in Livestock," California Agricultural Experiment Station Extension Service, University of California Circular 506, 1961.

10 Louis Klein: *River Pollution. 2: Causes and Effects*, Butterworth & Co. (Publishers), Ltd., London, 1962.

11 P. R. Gorham: *Biological Problems in Water Pollution*, U.S. Public Health Service Publication 999-WP-25, Robert A. Taft Sanitary Engineering Center, 1965.

12 C. M. Weiss: "Use of Fish To Detect Insecticides in Water," Publication 84, Department of Environmental Sciences and Engineering, University of North Carolina, May 5, 1964.

13 R. P. Kamrin and M. Singer: "The Influence of the Nerve on Regeneration and Maintenance of the Barbel of the Catfish, *Ameirus nebulosus*," *J. Morphol.*, 96(1), January, 1955.

14 *Proceedings, Conf. Physiological Aspects of Water Quality*, Washington, D.C., September 8-9, 1960 (available through U.S. Public Health Service, Division of Water Supply and Pollution Control).

15 C. Henderson, Q. H. Pickering, and J. M. Cohen: "The Toxicity of Synthetic Detergents and Soaps to Fish," *Sewage Ind. Wastes*, 31(3), 1959.

16 R. D. Swisher, J. T. O'Rourke, and H. D. Tomlinson: "Fish Bioassays of Linear Alkylate Sulfonates (LAS) and Intermediate Biodegradation Products," paper presented at 55th Annual Meeting, American Oil Chemical Society, New Orleans, April 20, 1964.

17 M. M. Ellis: "Detection and Measurement of Stream Pollution," *Bulletin of the U.S. Bureau of Fisheries*, 48(22), 1937.

18 *Water Pollution Research,* 1955; H. M. Stationery Office, London, 1956.

19 J. Grindley: "Toxicity to Rainbow Trout and Minnows of Some Substances Known To Be Present in Waste Water Discharged to Rivers," *Ann. Appl. Biol.,* 33, 1946.

20 State of Washington 64th Annual Report of Fisheries, 1944.

21 N. H. Sanborn: "The Lethal Effect of Certain Chemicals on Fresh Water Fish," *Canning Tr.,* 67, 1945.

22 J. R. E. Jones: "The Relation between the Electrolytic Solution Pressures of the Metals and Their Toxicity to the Stickleback (*Gasterosteus aculeatus L.*)," *J. Exp. Biol.,* 16, 1939.

23 P. Doudoroff and M. Katz: "Critical Review of Literature on the Toxicity of Industrial Wastes and Their Components to Fish. I. Alkalies, Acids and Inorganic Gases. II. Metals as Salts, *Sewage Ind. Wastes,* 22(11), 1950; 25(7), 1953.

24 F. L. Coventry, V. E. Shelford, and L. F. Miller: "The Conditioning of a Chloramine Treated Water Supply for Biological Purposes," *Ecology,* 16(1), 1935.

25 R. S. Taylor and M. C. James: "Treatment for Removal of Chlorine from City Water for Use in Aquaria," U.S. Bureau of Fisheries, Document 1,045: Report U.S. Commissioner of Fisheries app. 7, 1928.

26 *Water Pollution Research,* 1958; H. M. Stationery Office, London, 1959.

27 J. R. E. Jones: "The Relative Toxicity of Salts of Lead, Zinc and Copper to the Stickleback (*Gasterosteus aculeatus L.*) and the Effect of Calcium on the Toxicity of Lead and Zinc Salts," *J. Exp. Biol.,* 15, 1938.

28 L. A. Allen, N. Bleazard, and A. B. Wheatland: "Formation of Cyanogen Chloride during Chlorination of Certain Liquids: Toxicity of Such Liquids to Fish," *J. Hyg.,* 46, 1948.

29 K. E. Carpenter: "The Lethal Action of Soluble Metallic Salts on Fishes," *Brit. J. Exp. Biol.,* 4, 1927.

30 L. Van Dam: *On the Utilization of Oxygen and Regulation of Breathing in Some Aquatic Animals,* Drukkerij 'Volharding,' Groningen, 1938.

31 W. B. Alexander, B. A. Southgate, and R. Basindale: "Survey of the River Tees. Pt. II. The Estuary—Chemical and Biological," Technical Paper No. 5, Water Pollution Research, London, 1935.

32 J. R. E. Jones: "The Oxygen Consumption of *Gasterosteus aculeatus L.* in Toxic Solutions," *J. Exp. Biol.,* 23:298-311, 1947.

33 J. Longwell and F. T. K. Pentelow: "The Effect of Sewage on Brown Trout (*Salmo trutta L.*)," *J. Exp. Biol.,* 12:1-12, 1935.

34 R. Lloyd: "The Toxicity of Zinc Sulphate to Rainbow Trout," *Ann. Appl. Biol.,* 48:84-94, 1960.

35 C. A. Bond, R. H. Lewis, and J. L. Fryer: "Toxicity of Various Herbicidal Materials to Fishes," in *Biological Problems in Water Pollution, Trans. 1959 Seminar* Robert A. Taft Sanitary Engineering Center, Tech. Rept. W60-3.

36 J. Mayhew: "Toxicity of Seven Different Insecticides to Rainbow Trout, *Salmo gairdnerii Richardson*," *Proc. Iowa Acad. Sci.*, 62, 1955.

37 R. W. Hatch: "Relative Sensitivity of Salmonids to DDT," *Progr. Fish-Culturist*, 19:89-91, 1957.

38 D. I. Mount: "Chronic Effects of Endrin on Bluntnose Minnows and Guppies," Research Report 58, U.S. Fish and Wildlife Service, Bureau of Sport Fisheries and Wildlife, Washington, D.C., 1962.

39 C. Henderson, Q. H. Pickering, and C. M. Tarzwell: "The Toxicity of Organic Phosphorus and Chlorinated Hydrocarbon Insecticides to Fish," in *Biological Problems in Water Pollution, Trans. 1959 Seminar* Robert A. Taft Sanitary Engineering Center, Tech. Rept. W60-3.

40 *Water Pollution Research*, 1953; H. M. Stationery Office, London, 1954.

41 K. Latomi et al.: "Toxicity of Endrin to Fish," *Progr. Fish-Culturist*, 20, 1958.

42 E. Langor: "Pesticides: Minute Quantities Linked with Massive Fish Kills; Federal Policy Still Uncertain," *Science*, 144:35 1964.

43 J. E. Hemphill: "Toxaphene as a Fish Toxin," *Progr. Fish-Culturist*, 16, 1954.

44 Final Report, "Research Directed toward Development of Test Procedures for Evaluating Allowable Limits of Concentration of Toxic Substances in Aquatic Environments," Engineering-Science, Inc., Arcadia, Calif., November 30, 1962.

TABLE 13-10   *Water quality objectives, applicable to receiving waters, for salt and fresh surface waters and underground waters*

Water quality water uses	Organisms of the coliform group	Floating, suspended and settleable solids and sludge deposits	Taste- or odor- producing substances	Dissolved oxygen	pH
A. Water supply, drinking, culinary and food processing Without treatment other than simple disinfection and removal of naturally present impurities	Most probable number coliform bacterial content of a representative number of samples should average less than 50 per 100 ml in any month	None attributable to sewage, industrial wastes or other wastes, or which, after reasonable dilution and mixture with receiving waters, interfere with the best use of these waters for the purpose indicated	None attributable to sewage, industrial wastes, or other wastes	Greater than 5 parts per million except for underground waters	Hydrogen ion concentration expressed as pH should be maintained between 6.5 and 8.5
B. Water supply, drinking, culinary and food processing With treatment equal to coagulation, sedimentation, filtration, disinfection, and any additional treatment necessary for removing naturally present impurities	MPN coliform bacterial content when associated with domestic sewage of a representative number of samples should average less than 2000 per 100 ml and should not exceed this number in more than 20 percent of samples examined in any month	Same as for use *A* above	None attributable to sewage, industrial wastes, or other wastes which, after reasonable dilution and mixture, will increase the threshold odor number above 8	Greater than 5 parts per million except for underground waters	Same as for use *A* above

Toxic, colored or other deleterious substances	Phenolic compounds	Oil	High temperature wastes	Minimum treatment requirements for domestic sewage
None alone or in combination with other substances or wastes in sufficient amounts or of such nature as to make receiving water unsafe or unsuitable for use indicated (USPHS standards)	Less than 5 parts per billion	None	Not in sufficient quantities alone or in combination with other wastes to interfere with the use indicated	Sedimentation and effective disinfection
Same as for use *A* above	Less than 5 parts per billion	None alone or in combination with other substances or wastes as to make receiving water unfit or unsafe for the use indicated	Same as for use *A* above	Sedimentation and effective disinfection

TABLE 13-10 *Water quality
objectives, applicable to receiving
waters, for salt and fresh surface
waters and underground waters
(continued)*

Water quality water uses	Organisms of the coliform group	Floating, suspended and settleable solids and sludge deposits	Taste- or odor-producing substances	Dissolved oxygen	pH
C. Bathing, swimming and recreation Note: When waters are used for recreational purposes such as fishing and boating, exclusive of bathing and swimming, the number "1000" may be substituted for "240" in statement of coliform objective	Coliform bacterial content of a representative number of samples should average less than 240 per 100 ml and should not exceed this number in more than 20 percent of samples examined when associated with domestic sewage (see note under *C* at left)	Same as for use *A* above	None attributable to sewage, industrial wastes, or other wastes which, after reasonable dilution and mixture, will interfere with the best use of these waters for the purpose indicated	Greater than 5 parts per million	Same as for use *A* above
D. Growth and propagation of fish, shellfish and other aquatic life	Coliform bacterial content of a representative number of samples should not have a median concentration greater than 70 per 100 ml in waters used for the growth and propagation of shellfish	Same as for use *A* above	None attributable to sewage, industrial wastes, or other wastes which will interfere with the marketability or propagation of recreational or commercial fish, shellfish, or other edible aquatic forms	Greater than 6 parts per million	Same as for use *A* above

Toxic, colored or other deleterious substances	Phenolic compounds	Oil	High Temperature wastes	Minimum treatment requirements for domestic sewage
Same as for use *A* above	Less than 25 parts per billion or none in sufficient amounts such as to impart a residual taste to recreational or commercial fish, shellfish, or other aquatic forms	Same as for use *B* above	Same as for use *A* above	Sedimentation and effective disinfection
None alone or in combination with other substances or wastes to sufficient amount or of such character as to make receiving waters unsafe or unsuitable for use indicated	Same as for use *C* above	Same as for use *B* above	None in sufficient quantity as to be injurious to or interfere with the normal propagation of fish, shellfish, or other aquatic life	Sedimentation for all uses under this group, but disinfection required in addition only if discharged into waters used for the growth and propagation of shellfish, either commercial or recreational

TABLE 13-10   *Water quality objectives, applicable to receiving waters, for salt and fresh surface waters and underground waters (continued)*

Water quality water uses	Organisms of the coliform group	Floating, suspended and settleable solids and sludge deposits	Taste- or odor-producing substances	Dissolved oxygen	pH
E. *Agricultural and industrial water supply** Without treatment except for the removal of natural impurities to meet special quality requirements, other than those classified under *A* above		Same as for use *A* above	None attributable to sewage, industrial wastes, or other wastes which will adversely affect the marketability of agricultural or industrial produce	Greater than 3 parts per million	Hydrogen ion concentration expressed as pH should be maintained between 6.0 and 9.5

*For agricultural water supply, salinity and sodium hazards are determined by electrical conductivity ($EC \times 10^6$) and sodium adsorption ratio (SAR). Waters high in both salinity and sodium are generally unsuitable for irrigation purposes. (See "Classification and Use of Irrigation Waters," Circular No. 969, U.S. Department of Agriculture, November 1955.)

SOURCE: McKee and Wolf [6]. Reproduced by permission of the California State Water Control Board.

Toxic, colored or other deleterious substances	Phenolic compounds	Oil	High temperature wastes	Minimum treatment requirements for domestic sewage
Same as for use A above	None in sufficient quantity as to make receiving water unsuitable for use indicated	Same as for use B above	Same as for use A above	Sedimentation and effective disinfection

CHAPTER 14

Engineering aspects
of water quality management

INTRODUCTION

In preceding chapters attention has been directed to the concept of quality as a measurable dimension of water and to the interchange systems which affect the quality of water, either under primitive conditions or when diverted by man to various beneficial uses. Such relationships as diagramed in Chapters 3 and 8 serve to identify the bridges or critical points in these systems where specific engineered works, or subsystems, might be effective in changing the quality of water to meet the needs of users; or where the quality of water might be examined for conformity to requirements established by public policy. To a lesser degree some of the phenomena which might be exploited by engineered systems have been identified. The nature of these engineered works, however, depends upon such factors as the changes in water quality necessary to achieve accepted quality objectives, the resolving power of known processes, the status of current technology, and the compatibility of water quality management objectives and the economic realities of the times. More significant, perhaps, it depends upon the role the profession of engineering is prepared to play, or does in fact play, in the whole process of resources development in general, and, in particular, in the management of water resources for quality and other of our multipurpose objectives. Therefore, as a prelude to an analysis of either new or conventional methods of upgrading the quality of water, it is desirable to establish a clear picture of the role of engineering in water resources management and public works.

THE NATURE AND ROLE
OF PROFESSIONAL ENGINEERING

The nature of professional engineering practice has been described by M. P. O'Brien* by a categorical scale representing decreasing freedom

*Emeritus Dean of Engineering, University of California, Berkeley.

of decision on the part of the engineer. In the context of the development and management of water resources involving considerations of quantity and quality ultimately expressed in public works, the appropriate scale might be somewhat as follows:

1. Planning and policy development
2. Systems design and analysis
3. Process and functional design
4. Construction of physical works
5. Operation and maintenance of system and works
6. Equipment and process sales
7. Field service

Although the responsibility of the engineer for decision making varies from situation to situation, and may perhaps be a continuous rather than a discrete variable, the foregoing scale may be used to explain the various ways in which engineering is involved in water quality management.

Planning and policy development

The maximum freedom of decision exists, of course, when no decisions have yet been made and the problem is one of establishing the social objectives of resource development and management, and of formulating appropriate policies for their achievement. At this level the decision is made that "clean water," for example, is to be our national goal; or that water resources shall be developed by government for some spectrum of multiple purposes, and in some relationship to the private sector of the economy. In such a context the Tennessee Valley Authority was created; California decided to develop its own water resources; Arizona, in 1967, began to consider the feasibility of developing the Central Arizona Project as a state undertaking; and various groups initiated discussion of the desirability of vast interregional or international transfers of water in the West.

A full consideration of the decision process at the topmost level of freedom is, of course, beyond the scope of this work. Nevertheless, the role of the engineer in the process should be clearly understood. Ideally, the alternatives on which a final political decision is made in the field of water resources management should come from a team of experts from the several disciplines concerned, e.g., engineering, planning, public administration, economics, political science, law, finance, and legislation. Historically, however, no such team has been utilized, and one or more of its ideal components has played the dominant role. That one component, however, has not been the engineer, nor even the planner. These two, and other subdominant professions, have generally played the secondary role of advising on the technical feasibility of implementing decisions derived from various political considerations, motivations, or expedients. In such cases the role of the engineer at the planning and policy level is merely that of appraising the "fail-safe" features of any overall scheme.

This traditional secondary role of the engineer in top level decisions where resources and public works are involved has often been deplored by the engineering profession. Sometimes the public takes the engineer to task for what it considers his failure to assume full responsibility in public affairs. The implications in these complaints is often that engineering rather than some other profession should play the dominant role at the planning and policy level of decision making. Such a conclusion, though based on a certain logic is, however, hard to document because there is in fact a considerable degree of interchangeability of professions at the level of political decision. Each has its limitations as a fountainhead of policy decisions. The political scientists, for example, might propose something technically naive; the engineer, on the other hand, may develop alternatives which, although technically feasible and economically sound, are yet unacceptable to people for reasons varying from simple prejudice to complex political motivations. Of the two alternatives— to involve a wide spectrum of professions in a secondary role, or to include them all in a top level team—the latter is the most practical and defines the proper role of the engineer at the policy level of decision making.

Systems design

A decision that some level of government shall be responsible for developing water resources for specified multiple purposes in an area or region, when implemented by legislation held valid by the courts, eliminates only the freedom to decide on overall policy. There remains a very broad area of engineering concern which also involves considerations beyond technology and economics; hence, in deciding upon just what systems might achieve the policy objectives, a variety of disciplines or professions are again concerned, although the importance of the role of the engineer in the ultimate decision is increased. Examples of decisions at this level include the concept of the TVA system, the California Water Plan, the Texas Water Plan, and similar complex and broad blueprints for feasible action. Here alternate systems of interbasin transfer, storage, conjunctive use of surface and ground waters, flood routing, etc., are set forth and mathematical models sought for analyzing entire systems in relation to economics, physical operation, and staging of components.

In the realm of water quality management, such problems as salt buildup in stored water; the effect of agricultural, domestic, and industrial return flows on ground- and surface-water resources; and the effect on water quality of allocating water to various sectors of the economy might be considered in relation to alternate systems of water management in order to minimize the cost of engineered works specifically applied to water quality control.

Historically, engineering decisions at the systems design level have been the result of judgment based on far less sophisticated tools than the computer and with minimal concern for water quality. With the advent of modern systems-analysis techniques and the growing difficulty of achieving water quality goals under expanding resource use, the role of the engineer at the systems design level of decision is increasing rapidly both in importance and in complexity.

Process and functional design

It is at this level of decision making that the engineer plays what is perhaps his most accustomed role. Here he is no longer free to decide what public policy would be most sound, whether water from one region should be delivered to another, what role the sale of power should play in project financing, or similar policy matters. Prior decision has established the need for dams, powerhouses, canals, and other physical structures. Nevertheless, important decisions must be made concerning location of structures, types of dams, cost of alternate works, etc., in considering how the project is to be done most efficiently and economically within the prescribed constraints. Thus a concern for functional design is an inherent part of water resources development. As water quality considerations become constraints on functional design, however, attention must be given to the way in which alternate arrangements of "hardware" affect functional design. For example, instances of degradation of water quality in a stream have been related to low-elevation intake of stored water to hydraulic turbines.

In situations where engineered works to alter the quality of water are a part of the overall system, process design must precede some aspects of functional design. Thus engineering decisions must be made concerning the sequence of unit processes appropriate to the required change in water quality, as well as the type of structure most suited to the process.

Construction of physical works

Obviously, once plans and specifications have resulted from decisions of the process and functional designers, to build or not to build is no longer a question for the engineer to decide. The question now is how to build the project within the economic constraints agreed upon in contractual documents. Here the decisions are largely made by engineers working with associates in their own organizations; and from the standpoint of freedom of decision, the latitude is quite narrow at this level. Nevertheless, important decisions of a thoroughly professional nature must be made by the construction engineer, for on him depends the success of the project. His decisions, however, are no longer directly related to water quality management per se unless they should reveal some factor overlooked by the designer which might be corrected prior to construction.

Operation and maintenance
of system and works

Once the structures and hardware which implement a system are in being, what it produces becomes a function both of its inherent capabilities and how it is operated and maintained. In a complex system such as the TVA, for example, there is an extremely large area of engineering decisions which affect its functioning. Some of these decisions relate directly to water quality and may require the weighing of quality maintenance against some other purpose which might be given first preference. Although from the viewpoint of a single project the area of engineering decision is less at the operational level than at the construction level, such may not in reality prove to be the case. A whole new set of policy decisions based on current resource concepts and proposed or accepted new systems may be introduced at the operational level of any project by a line of decisions which parallel but do not retrace the pattern which initially created that project. As a hypothetical example, the operational relationships of the state-owned California Water System and the federally owned Central Valley Project might eventually be completely recast by policy decisions leading to interbasin transfers of water in the western United States and Canada.

Equipment and
process sales

At the level of equipment and process sales the latitude for engineering decision becomes quite narrow as a result of previous decisions by others, both on the project and in the equipment manufacturing company. Advice by the sales engineer may be invaluable to the functional and process designers, to the construction engineers, and to the engineers in charge of maintenance and operation in the course of making decisions in their respective areas of responsibility, but there is little room for him to make independent decisions. Basically he may decide whether the product his company can furnish is suited to the application under the conditions proposed for loading, exposure, and so forth; or which type and size of equipment manufactured by his company are to be recommended; or which scientific principle exploited by his alternate devices is most economical, effective, and reliable for the installation. The final decision as to whether one process or another is most suited to the water quality or other management goal of the project, however, goes back to the process designer for final decision. Nevertheless, in spite of a limited freedom of decision at the sales level of engineering, the decisions themselves require engineering judgment of a thoroughly professional degree.

Field service

Although requiring engineering competence, field service offers the engineer little latitude for decision making. Problems of installing equipment in a specific situation, however, may require a high order of engineering judgment on the part of the service engineer. For example, equipment checked out at the factory in summer has been known to defy assembly when delivered to a job site in winter. Both expense to the manufacturer and loss of confidence in his product by the purchaser have then been prevented by a correct analysis of the reasons for failure, and intelligent decisions on how to overcome them, by the service engineer. In other situations the engineer on field service may be called upon to decide whether his company or the purchaser is responsible for failure or malfunction of equipment within a warranty period, and perhaps, whether it is in the interests of his company to accept responsibility for faults of the purchaser in a particular instance of malfunction.

PREPARATION
OF THE ENGINEER
FOR HIS ROLE

If in the interests of sound water quality management the engineer is to participate in decision making at several levels ranging from the purely technical to those involving social goals of modern society, it is important that his preparation for such a role be significant. Although an analysis of the pros and cons of engineering education is beyond the scope of this chapter, some facets of the engineer's background are pertinent to his role in resources development and management. Of primary importance is the fact that in 1965 some 80 percent of professional engineering practice was included in the scale range from "process and functional design" to "field service." The other 20 percent was in the top two categories—planning and systems. Because of this fact the objectives of engineering colleges have differed markedly in the past—some, predominantly the undergraduate schools, seeing their role as that of educating for the biggest sector of demand; others, primarily the graduate schools, looking to planning and systems as their major educational goal.

The natural tendency, however, is for each to aspire to equal the best. The result is a definite shift upward in the scale, leaving behind all preparation for the nonprofessional aspects of engineering and creeping up the list of professional categories until functional design may be all but eliminated; essentially no formal preparation at all is directed to the construction, operation, sales, and service aspects of professional engineering.

In this educational climate the engineer is being well prepared in his technical studies for the type of systems design and analysis required for resources development and management on a broad scale. Thus the next generation of engineers may be expected to play a more active role at this level of decision. Formal preparation for planning and policy development is not so straightforward a problem. There are no courses in "decision making" and no demand for decision makers fresh from the Ph.D. program. Here judgment born of experience, and development of knowledge in such areas as economics, public administration, planning, and the various humanities, enables the engineer to carry his engineering know-how into the councils of the decision-making team. Without his engineering background, he has nothing particular to contribute; and without the broader educational background he may never have the opportunity to participate or to communicate with others described in a previous section as "interchangeable" at the top level of decision. In recent years the preparation of the engineer for the top level of decision making has become increasingly sound but is still in flux.

ECONOMICS IN ENGINEERING

Second only to considerations of physical integrity and safety of the structures he designs is the engineer's concern for economy. Essentially all definitions of engineering list lowest cost among the profession's manifest benefits to society. In fact, the engineer tends to be so cost conscious that he is often accused of violating the simplest principles of aesthetics in the interest of economy. Regardless of the validity of such an accusation there are several reasons why the engineer concerned with water quality management should be conscious of costs. First, projects directed to quality objectives are typically water or waste-water treatment plants for individual cities, districts, or industries. City and district projects, of course, represent public works at the local level where competition for tax dollars is notably acute. Here low cost is often so attractive to public officials that one of the problems of engineering is to protect the community, particularly the small one, from unethical individuals willing to put cost above sound engineering. A second factor leading to economics as a factor in engineering for public works at the local level is the public attitude toward wastes. However unhappy the public may be with polluted water as an aspect of the environment, it takes an extremely conservative view of what it can "afford" to pay for renovation of its wastes. Reflecting this attitude, elected officials understandably consider waste treatment plants among the least monumental of public works that must be provided.

In the case of individual industries, pressure for economy in engineered works for water quality management is possibly more compelling than in public works. Here competition for markets and profits dictates that investment in nonproductive activities be kept at a minimum. Minimal cost of process and cooling water is a factor in industrial plant location. The quality of waste water which the plant may discharge, however, cannot be established for all time in advance. Public policies related to water quality objectives change with time, and as noted in previous chapters, are becoming increasingly restrictive. Thus the individual industrial plant is confronted with a growing problem of waste treatment as a part of its cost of doing business; this in turn is reflected in pressure on the engineer to provide waste-water treatment for industry at the least possible cost.

While the place of economics in engineering for water quality management is quite clearly understood at the local level, it is far less clearly defined, and indeed unresolved, at the national, regional, and state levels of water resource development. This is because quality management has not been one of the major considerations in water resource development and management in the past to the extent that it is destined to be in the future. Although resource economics constitutes a whole subject in itself which is far beyond the scope of this work, brief consideration of the problem may point to the role of economics in engineering for water quality management in the years ahead.

Several factors serve to make impossible a simple extrapolation from the local engineered subsystem to the overall system involved in water resource development and management. For one thing, the project is farther removed from the individual taxpayer who either sees unclearly his interest in public economy at higher levels, or, resigned to the inevitability of taxes, is anxious that his "share" be spent on projects in his immediate area of interest rather than elsewhere in the nation or region. While this does not relieve the engineer of concern for economy in his planning, it does move economic factors to those levels of decision where engineering is not the sole determinant of project costs. This shift is further directed by the difference in purpose of the project and hence in the economic parameters to be considered. Specifically, the local waste treatment plant is designed with the objective of changing the quality factors in a given amount of water from some analytically observable spectrum to some other such spectrum demanded by regulatory agencies pursuant to stated public policy objectives. The resource development project, on the other hand, involves a wide variety of purposes not all of which are readily evaluated. Furthermore, water quality management has traditionally been one of the less clearly defined purposes, hence there is minimal experience in its evaluation.

The problem of economics in water quality management may be underscored by considering a hypothetical example of a multipurpose project involving such beneficial uses as flood control, power production, navigation, irrigation, water supply, recreation, fish and wildlife preservation. and waste dilution. Although as noted in previous chapters the tendency of public policy today is to drop waste-water dilution from the list of beneficial uses—and water quality management is not specifically listed—quality is an implicit consideration in all but the first three beneficial uses listed.

On the basis of past experience we might expect this project to be analyzed economically on a cost-benefit basis. In such a case current or foreseeable cost would be computed for the system in a normal engineering economics fashion, subject to such variations in interest rates and financial costs as set forth in public policy. Against this project cost the following would be compared: the estimated value of flood control; the value of power to be sold in accord with public policy; the return of costs by irrigators and other purchasers of water, again in line with public policy; the value to navigation; and any other source of income to the project. In the ideal case an excess of these benefits over the project cost would justify the project, and such values as recreation and wildlife preservation could be considered as added positive values without assigned numbers. In the less than ideal case, however, an unfavorable cost-benefit ratio might not preclude a decision to construct the project. Arbitrary assignment of numerical values to recreation and wildlife, reduction in salinity of streams, future changes in the pattern of water sales, and other policy matters may lead to favorable political decision quite aside from traditional concepts of engineering economics. While this may lead opponents of the project to cry "politics," the important fact here is that economics, in the sense to which engineers are accustomed in their activities, may or may not be the determining factor in water resource development decisions. For the future a whole new set of decision parameters involving both engineering and resource economics decisions could result from a series of developments ranging from a reassessment of the value of flood control if occupation of the flood plain leads to damage by only minor floods, to the effect of nuclear power on the cost-benefit ratio of water resource development projects.

In the foregoing hypothetical example of a water resources management project it is notable that management for water quality control objectives is not directly introduced. For the future, quality considerations are involved in direct (water supply and irrigation) benefits, and in indirect (recreation and wildlife) benefits. Considerations presented in preceding chapters show that quality is involved in each of these benefits in a different manner and to a different degree. Furthermore, it is involved in water storage and in flood control procedures. How is the engineer concerned with water quality management at the resource development level to apply the concepts of economics which he routinely applies to the design, construction, and operating levels of decision? This is a very real question and points up an increasingly complex role of economics in engineering.

INNOVATION IN ENGINEERING

Development of ingenious, novel, and imaginative ways to accomplish useful things for mankind at a low cost and with all but absolute reliability is a part of the credo of the engineer. Unfortunately it does not always characterize engineered works. Like decision making, ingenuity and imagination cannot become attributes of every individual successfully completing degree requirements in engineering. They are, however, common attributes of the engineering profession and characterize any highly successful construction firm. Freedom to employ innovation might be expressed by a categorical scale somewhat like but not paralleling the previously noted scale of decision making. Here the construction engineer would be near the top of the list since freedom to innovate largely describes his freedom of decision. Lower in the scale might be the functional and process designer whose task it is to meet resource-oriented water quality objectives by engineered works at the critical bridges in the systems depicted in Chapters 3 and 8.

In this context the factors which make for an acute awareness of economics in local public works make engineering innovation all but impossible. Municipal officials simply cannot risk acceptance of a system which is not customarily being applied elsewhere. Conversely, the system may be acceptable if it is customarily used, even though it does a poor job. This, together with what seems to be an innate conservatism in public works engineers, accounts in part for a dismaying lack of progress in water treatment in more than a generation. Attention is therefore directed in this and subsequent chapters to the resolving power of conventional methods of quality management and to some less utilized possibilities.

THE ART VERSUS THE SCIENCE OF WATER TREATMENT

Conventional methods of purifying water for beneficial use, or of upgrading the quality of its return flows, to a large degree developed as an art rather than as a science. That is to say, engineers knew how to accomplish certain results long before anyone fully understood how the system functioned. Although this is a tribute to the ingenuity of man, it does not overcome the fact that an art often tends to adhere to custom, changing by evolution rather than by innovation. Thus, for a long time engineered systems for changing water quality tended to be stereotyped devices rather than unique solutions to a specific problem. Within this narrow concept unpotable water was made potable by the construction of a filtration plant. That the plant improved somewhat from generation to generation is beside the point—a filtration plant was an object offered to the water-consuming community much as a chair might be the answer to a traveler's weariness.

In more recent years, however, human curiosity, advancing sanitary standards, needs of a burgeoning population, expanding knowledge, and the cult of science have had their effects on water quality management. The change, while profound, cannot be said to be revolutionary because more than a point of view is involved. The extremely complex relationships of biochemistry and the intricate equilibria of nature are but slowly revealed through research. The result is that today a very large gap still remains between what is known scientifically and what is done practically in water treatment systems. In many cases it is still too early to begin with basic scientific principles, apply experimentally determined coefficients, and come up with criteria for design. Furthermore, the rapid advance of knowledge tends to isolate the researcher from the practicing engineer. Thus the spectrum of knowledge held by the designer may not encompass all that mankind knows of the problem at hand. This deficiency, together with the comfort that goes with accustomed things, a traditional lack of freedom to innovate in public works, and economic constraints on the problem, tends to perpetuate obsolescent designs. However, the science of water quality control is beginning to have an impact on the art, and the better engineers today look upon water treatment as a problem of determining what processes are necessary to change the observed quality of a water to some other quality appropriate to its use or to the objectives of water quality control. Thus a unique solution based on scientific knowledge and current technology becomes the engineer's answer to water quality management.

INADEQUACY OF
TRADITIONAL CONCEPTS

Although the engineering feasibility of putting theory into practice is increasing from year to year, the problems of water quality management generated by the normal cycle of growth and decay of organic matter alone still present a more than adequate technological challenge. Meanwhile the rise of industry as the major producer of wastes and the widespread commercial use of industrial products have produced problems completely beyond the range and the resolving power of conventional engineered systems. Particular attention has already been called in preceding chapters to the stable end products of organic degradation—nitrates and phosphates—and to such materials as pesticide residues. To cope with some of these products by removing them from water now calls for the development of systems based on scientific principles and the application of results of research, rather than for a simple modification of ancient arts. In some cases it may also suggest a change in the concept of wastes management.

One of the most traditional of concepts is that of treating the entire volume of water in order to remove small amounts of undesirable materials. For the purpose of putting water to beneficial use, once the unwanted quality factors were present, there was no alternative to this approach. However, this same concept carried over to the return streams of beneficial use. With the invention of the water closet, the sanitary sewer was conceived as an engineered device to transport wastes beyond the confines of the community for discharge into some stream which served as a natural transport system. Later, when concern for quality of the stream made necessary the upgrading of domestic return waters, there was little choice but to handle the entire volume of waste-water flow to search out the quality factors which were to be removed or stabilized. Logically this same concept was extended to industrial return flows. Here, however, the concept proved inadequate when industrial wastes were toxic to biological treatment systems or were immune to biodegradation. Thus began a concept of engineered systems applied by industry to its own waste streams, or of inplant changes for the purpose of excluding certain materials from the return flows. Nevertheless, the general concept of waste-water treatment for control of quality of the water resource is still heavily oriented to the cycle of organic degradation and thus inappropriate to the range of materials it seeks to control; and the parameters used to judge its performance are too limited to detect its effect on a host of more subtle quality factors.

ORIENTATION
TO MECHANIZED SYSTEMS

For reasons that are obvious to any citizen, the development of waste-treatment systems in the United States has taken the direction of increasingly mechanized, instrument-controlled plants into which waste streams flow and from which they emerge sufficiently altered in quality to meet requirements imposed by regulatory agencies. Because such systems involve maintenance of mass cultures of organisms (e.g., activated sludge) or require a whole sequence of environments to exist simultaneously in a perilous balance (e.g., sludge digestion), they require the attention of highly capable engineers and scientists. This leads inevitably to the mistaking of sophistication of mechanization for sophistication of process, particularly in contrast with such natural systems as the soil mantle of the earth or the surface pond in which a very much more complex biological system functions without man's assistance. The effect is worldwide and in many ways unfortunate. The industrially undeveloped country looking to the West for the technology which makes for economic growth accepts the mechanized waste treatment plant as a symbol of equal validity with the steel mill. Thus both inadequate engineering leadership and a desire to be unassociated with primitive and Stone Age systems have prevented the unmechanized country from developing natural systems to a high degree. Preoccupation with mechanized systems has had a similar effect in the United States with the result that, as noted in Chapters 16 and 17, attention has only recently been directed to the potential of the soil and the pond to function either as engineered systems for primary or, in concert with conventional mechanized systems, for secondary or tertiary treatment of return flows from beneficial uses.

ENGINEERING APPROACHES

There are two principal types of approach to engineered systems for water quality management:

1. Removal of water from waste-water return flows (water reclamation)
2. Removal or stabilization of the quality factors in water (water renovation)

The two are distinguished more by the objective of the project than by the processes utilized to accomplish that objective. In general, reclamation suggests a recycling of water to reduce the demand of beneficial use or to avoid the probem of pollution of the water resource by its return flows. Water renovation, on the other hand, implies an upgrading of quality of water either for beneficial use or for release to the water resource. Needless to say, the latter approach is the most widely used. It is commonly known as water treatment, or purification; sewage treatment, or purification; waste-water treatment; and similar generic terms for a squence of processes deemed by the engineer to be appropriate to the water quality control task at hand. In quite recent years the term "renovation" has come into use to describe the concept.

Water reclamation generally implies that an appreciable liquid residue remains to be renovated or conveyed to the ocean sink without mingling with any important sector of the water resource. Water renovation, on the other hand, implies that the entire mass of water is rendered useful. This concept, however, is valid only in the traditional circumstance of domestic waste-water treatment in which solids are removed for land disposal or left in the water as stable compounds. In cases where removal of nutrients, minerals, and various ions is necessary for quality management, the disposal of residues likewise poses a problem.

In the chapters which follow, engineering approaches to water reclamation are considered in some detail.

CHAPTER 15

Water reclamation

INTRODUCTION

The term "reclamation" has long been associated with water management activities, although in the historic sense it is the land that is to be reclaimed by irrigation water. More recently, attention has been turned to the recovery of water itself—which has been degraded in quality to the extent that it is no longer suited to beneficial use, yet too valuable to throw away in many a water-deficient area. The term "water reclamation" has therefore come into quite wide use although without precise definition. In some cases it has been considered as synonymous with reuse of water; in others, its intended meaning is more accurately described as water renovation. For the purpose of this discussion, water reclamation is considered as a distinct objective of engineered systems, although the processes by which it is accomplished may involve renovation techniques.

SOURCES FROM WHICH WATER MAY BE RECLAIMED

The sources from which water might be reclaimed include:
1. Oceans and estuaries
2. Saline groundwaters
3. Agricultural return waters
4. Industrial return waters.
5. Domestic return waters

In some of these cases the need is to desalt waters which have acquired by dissolution a solids content too great for most beneficial uses. In others, it is desired to recover water which has acquired undesirable quality factors during use as a transportation device intended to remove wastes from the environment of their origin. Generally, only domestic and industrial return waters fall into this latter category, although in some instances desalination of domestic waste water would be required to renovate it. For the purposes of this discussion, domestic and industrial wastes are considered to be combined and are hence hereinafter called municipal waste water or return flow. This reflects the fact that domestic waste water in any reclaimable amount is seldom separate from industrial inputs. On the other hand, reclamation of water from industrial flows alone is generally an inplant operation which may or may not be directly related to water quality management objectives.

METHODS OF RECLAIMING WATER

There are several methods of reclaiming water from the various possible sources. Currently most of the effort to reclaim mineralized water is directed to desalting of seawater, with some attention going to saline groundwaters and essentially none at all to irrigation return waters. Seawater conversion, although a reclamation procedure of great importance, has never been motivated by resource quality management objectives; and irrigation return flows, although a critical factor in water quality, are not currently the subject of reclamation undertakings. Consequently, neither is herein discussed in detail. Instead, attention is directed to municipal return flows.

Methods of reclaiming water from municipal return flows include:

1. Especially designed reclamation plants involving biological treatment, filtration, chlorination, etc.
2. Use of algal ponds in which nutrients in waste waters are converted to separable algal cells
3. Conventional sewage treatment followed by partial deionization to remove the increment of mineral content added by use
4. Engineered soil systems
5. Conventional secondary sewage treatment plants.

Of these five methods the last has been most used simply because the idea of reclamation grew out of the problem of disposal of sewage effluent where no receiving water existed, together with the critical need for water in some parts of the semiarid Southwest. Therefore, it is more associated with incidental reclamation than with purposeful engineered water reclamation enterprise, and the effluent produced is not optimum for reuse.

The first four methods, although particularly suited to reclaiming water, are in various stages of development. Especially designed reclamation plants are in operation at Whittier Narrows [1] and Golden Gate Park in California. Algal ponds especially designed for nutrient removal have been developed at the University of California [2, 3]. However, the economics of harvesting of algae are not yet well established. The partial-demineralizing technique has recently been reported by Sanks and Kaufman [4, 5] and appears to have good possibilities. Engineered soil systems are currently limited to sand filters; however, the scientific knowledge necessary for their design is rapidly emerging [7].

DETERRENTS TO PURPOSEFUL WATER RECLAMATION

For the most part the reclamation of water from municipal wastes has been incidental rather than purposeful [7]. That is, in the process of complying with effluent or receiving water standards, return flows have been so upgraded in quality that when mingled with the receiving waters they have been incidentally subjected to use more than one time. The approach, however, has been one of disposal rather than of water reclamation.

There have been numerous deterrents to purposeful water reclamation in the past. Among the most significant are:

1. The concept of waste treatment as a device for meeting the requirements of regulatory agencies concerned with the public health or with water pollution problems, rather than as a measure in conserving or managing water resources
2. Association of reclamation with disposal of waste effluents
3. Unclear concept of which beneficial use is to utilize reclaimed water
4. Overrestrictive legislation predicated upon assumptions of pollution travel
5. Preoccupation with the needs for agricultural water which are far beyond the quantity of waste water available
6. Inertia—a tendency to persist in accustomed habit patterns
7. Tradition of wastefulness and rejection of secondhand or other unwanted materials
8. Cheapness of freshwater (approximately 5 cents per ton, delivered)
9. Uncertainty of the role of detergents, exotic organics, virus, etc., in the quality of reclaimed water
10. Uncertain economics of reclamation
11. Question of rights to groundwater replenished at public expense

FUNDAMENTALS OF PURPOSEFUL RECLAMATION

A number of the drawbacks to direct reuse of water outlined above are inherent in the concept of reuse as a method of disposing of sewage treatment-plant effluents. The first step in purposeful water reclamation, therefore, is to separate the sewage-disposal and water-reclamation functions of a system. The normal system of disposal is based on health and aesthetic considerations, hence it cannot be tied to continual acceptance of effluent by beneficial uses. That is, there is no way in which the duty of.public agencies to protect the public health can be subcontracted to private users through agreements requiring them to take all the effluent at all times. Separation of the two functions can be attained in several ways:

1. Reclaiming water from the sewer at the site of its use, at rates and at times suited to the user, and returning the effluents of the reclamation plant to the sewer
2. Reclaiming water from the sewer prior to introduction of industrial wastes which might preclude reclamation
3. Reclaiming water from domestic wastes at convenient points in the system for recharge of groundwaters

PRACTICAL CONSIDERATIONS IN RECLAMATION

Parkhurst [1], on the basis of field experience in water reclamation, states that four conditions are necessary to justify the construction of separate facilities for the purpose of water reclamation:

1. The chemical quality of the water must be suitable for reuse.
2. The quantity available must be sufficient to permit economical production costs.
3. Reclaimable water must be produced near a project which can utilize it.
4. A benefit must be derived from the project to provide interest in the purchase of water at a price to compensate for all or a part of the cost of production.

Concerning the factor of chemical quality, Parkhurst points out that the increment added by domestic use in terms of total dissolved solids is about 300 mg/liter. Assuming 1,000 mg/liter TDS as the limit of useful water, as is customary today, this means that raw water supplies may not contain more than 700 mg/liter TDS if the water is to be reclaimed by methods other than desalination after domestic usage.

Applying this criterion to waters in the Los Angeles Basin, Parkhurst's findings concerning reclaimable water in the area may be summarized as follows:

1. *City of Los Angeles*
   83 percent of water from Owens Valley with TDS of 200 mg/liter.
   17 percent of water from Colorado River with TDS of 600 to 800 mg/liter.
   Approximately 95 percent of nearly 300 mgd is reclaimable.
2. *Los Angeles County*
   70 percent of water from Colorado River, plus heavy industrial contributions of brine discharges.

Hence only 33 percent of some 285 mgd (i.e., 100 mgd) is reclaimable. 90 percent of the reclaimable water, however, could be put to practical use—aboveground reuse, or recharged to the groundwater.

Thus it seems evident that until the economics and technology of partial demineralization lead to practical application of the process, water reclamation from municipal wastes will continue to be practical only where salinity is not the controlling factor.

USES OF RECLAIMED WATER

Deterrents to widespread practice of water reclamation from municipal return flows have to a great extent involved uncertainties concerning the utilization of the product. Suggested uses of water reclaimed from this source include:

1. Direct reuse aboveground by
   (a) Industry
   (b) Agriculture
   (c) Recreational activities
2. Groundwater recharge by
   (a) Surface spreading
   (b) Direct injection into aquifers
   (c) Overirrigation of crops
   (d) Subsurface percolation systems

From the standpoint of economics, direct reuse aboveground would seem the most feasible; and of the three possible uses, *industry* is the most logical, if only because industrial activity is an intimate part of any municipality large enough to produce waste water in important amounts.

Since 60 to 80 percent of industrial water is used for cooling, treated domestic waste water might be used for this purpose with only slime control. However, in this case all the deterrents previously noted apply, plus some which are peculiar to the problem of industrial use. For example, delivery of water from the waste-water treatment plant to the industrial area of the city involves installing a separate system of pipes in paved streets already underlain with a network of conduits. In addition, the maintenance of two separate piping systems within an industrial plant is a nuisance and is hard to police by health agencies concerned with cross-connections.

It is from such considerations that it becomes evident that departure from the concept of utilization of treatment plant effluents is an important requirement if industrial use of reclaimed water is to be practical. However, special reclamation plants located at industrial sites and operated in accord with the needs of the user are an entirely feasible prospect. The effect of this, from a water quality management viewpoint, would be to increase the concentration of impurities in the residual flow. In an individual case this would have to be balanced off against the effect of industrial use of an alternate water supply on the quality of available water resources.

Direct reuse by *agriculture* and other land areas is not normally a good prospect, both for reasons of quantity and for reasons of geography.

Sewage effluents have in a number of cases been applied to crop and grass lands successfully. Instances of irrigation of pasture land, cotton, citrus fruits, grains, golf courses, etc., are noted in the literature [7]. Generally in these cases the motive has been disposal of waste water where no receiving water exists, or an extremely acute water shortage in the immediate locale. The major drawbacks to agricultural use are that crop type is limited if the farmer must take effluent on a year-round basis at a constant rate rather than on a seasonal basis as required by crops. Thus disposal of waste water by the community cannot be tied to agricultural use. More important, agriculture activity is generally located both uphill and remote from any important volume of reclaimable municipal waste water. And finally, the total quantity of water in domestic use, although impressive in amount, is but a few percent of agriculture's projected needs. For example, the approximately 1,000 mgd of domestic return water discharged to the ocean each day in California is only about 6 percent of the estimated need of agriculture for new water in the state.

Recent developments in recreational use of reclaimed water indicate increasing possibilities.

Planned reclamation of water from municipal wastes for recreational use has been demonstrated and practiced in Golden Gate Park at San Francisco for almost a generation. Here planted areas and a series of ponds are supplied directly with reclaimed water. At Las Vegas and several other places in the Southwest, golf courses are watered with sewage effluents. At the Santee Project near San Diego, recreational use of reclaimed water after passing through a soil system is well established. Because of concern for viruses and exotic organics as factors in the safety of reclaimed water, there is an increasing tendency to require passage through a soil system before using water for recreation. This may well place a limitation on the processes necessary to prepare water reclaimed from municipal wastes for direct reuse in recreational activities, but, as noted in Chapter 16, an engineered soil system is a feasible process.

Groundwater recharge is rapidly becoming the most popular method of utilizing water reclaimed from municipal return flows. Where suitable land area is available and underground storage or transport capacity exists, the method is limited principally by practical economic and jurisdictional consideration. From experience with groundwater recharge have emerged some of the concepts and parameters of engineered soil systems discussed in Chapter 16. However, as a method for water reclamation, factors such as technical feasibility, danger of pollution travel, practical rates, and operational procedures have been given first consideration from an engineering and water quality viewpoint. Nevertheless, a growing understanding of how to take advantage of the biologically active mantle of the earth in upgrading the quality of waste waters indicates that groundwater recharge can be made a reclamation technique, as well as a use of reclaimed water.

1. *Technical feasibility* Goudey [8] first demonstrated in 1930 that a highly treated sewage effluent (activated sludge followed by sand filtration) could be returned to the groundwater by surface spreading. In 1949, Arnold, Hedger, and Rawn [9, 10] demonstrated at Azusa and Whittier, California, that groundwater recharge with sewage effluents was practical. Later studies at Lodi and Bakersfield [11, 12, 13] developed the operational and management techniques for recharge with sewage effluents and floodwaters. In 1954 [14] similar parameters were developed at Richmond and Manhattan Beach for direct injection of water into underground water-bearing formations.

These and many other studies (see [7]) of a similar nature demonstrated the technical feasibility of groundwater recharge by surface spreading and direct injection. Overirrigation, a variant of surface spreading, has been generally eliminated as a major practical approach because of the seasonal nature of crop irrigation and the difficulty of changing the habits of irrigators.

2. *Travel of pollution* As previously noted, legal restraints upon groundwater recharge with domestic wastewater effluents were predicated upon the possibility that bacteria would travel underground with percolating and moving groundwater, in spite of the general evidence to the contrary. Random evidence of limited movement of bacteria and long-distance travel of chemicals was to be found in the literature [7] when systematic investigations of pollution travel were initiated in California in 1950. At that time experiments [7, 11] directed to both the public health and engineering aspects of surface spreading of sewage effluents were initiated at Lodi, and continued for 28 months. Previous experiments [10] at Whittier had shown a reduction from 110,000 coliforms per 100 ml to 40,000 per 100 ml in 3 ft of soil, with none appearing at lower horizons. In a coarser soil at Azusa, 120,000 coliforms per 100 ml was reduced to 60 at 2½ and 7 ft below the surface. At Lodi the observed penetration of coliforms was as shown in Table 15-1.

In all cases bacteria behaved as particulate matter, penetrating to a maximum depth in the soil initially and regressing as surface clogging developed and organisms died.

Such experimental data led to the conclusion that bacterial travel as a result of surface spreading of sewage treatment-plant effluents was not a factor for concern from the standpoint of water quality. Later

TABLE 15-1 *MPN of coliform
organisms as a function of depth
in Hanford fine sandy loam
at Lodi, California*

Basin	Sewage effluent spread	Surface	1 ft	2 ft	4 ft	7 ft	10 ft	13 ft
A	Primary	414×10^4	1.6	32*	0.6	0	0	
	Final	179×10^3	1.2	285*	2.1	0	0	
B	Primary	570×10^4	20	0	0	0	0	0
	Final	188×10^3	482	5.6	0.5	0.2	0.1	0
C	Final	188×10^3	148	305	2.0	0.2	0.1	0.3
D	Final	164×10^3	0.2	...	...	0	...	0

*Sand channel from surface to 2-ft depth.

(1965) observations [15] of movement of bacteria in a field-scale recharge operation [1] at Whittier Narrows fully confirmed this conclusion.

When sewage effluents were injected directly into an underground aquifer at Richmond [14], the distance of travel under quite high wellhead pressure was as shown in Table 15-2.

These data, together with those of other researchers, demonstrated quite conclusively that the danger of travel of bacteria with percolating water or moving groundwaters in a soil or sand medium is not a deterrent to the reclamation of waste water by groundwater recharge.

In the case of chemicals it was found that nutrients, whether stable or unstable, move quite readily with percolating water once they escape or are injected beyond the biologically active zone. Phosphates and ammonia tend to be adsorbed on soils, but nitrates and other ions normal to groundwater are not removed, and many inorganic and organic ions, ABS, and other chemicals move with groundwaters.

3. *The problem of virus* Demonstrations of the removal of bacteria in soil systems did not result in profound changes in the laws concerning the quality of water suited to recharge, although by 1965 1,000 coliforms per 100 ml began to be acceptable. The reason, beyond the innate conservation of health departments, was the unknown behavior of viruses in soils.

Enteroviruses found in sewage [16] include more than 60 types belonging to three groups:

(a)   The polio viruses

(b)   The ECHO viruses

(c)   The Coxsackie viruses

Adenoviruses and reoviruses, clinically considered respiratory, are also found in sewage. The virus of hepatitis must be presumed to be present, although propagation of this virus in the laboratory has not been confirmed.

Significant data on the removal of viruses by soil systems have now been produced at the Santee Project in California, in which the safety of waters reclaimed from domestic wastes is being demonstrated.

TABLE 15-2   *MPN of coliform organisms in observation wells during continuous recharge with an average of 2.4 × 10⁶ organisms per 100 ml*

Distance from recharge well, ft	MPN, 3rd day	MPN, 12th day	MPN, 32nd day
13 N	240	240,000	230
28 N	2400	240	5
47 N	240	38	5
63 N	23	8.8	None
88 N	None	None	None
138 N	. . .	None	None
39 NE	2400	240	8.8
45 NE	None	8.8	None
63 NE	None	38	None
106 NE	None	None	None
39 NW	2400	240	2300
45 NW	240	None	5
63 NW	None	2.2	8.8
13 E	24,000	24,000	8.8
50 E	240	5.0	None
13 W	23	None	2300
50 W	23	> 240	2.2
13 S	95	2400	230
63 S	None	None	9.4
100 S	23	5.0	None
188 S	None	None	None
192 S	None	None	None

At the Santee Project, it might be noted, domestic waste water is being renovated and used successfully for recreational purposes under the careful management and study of various state and national agencies. Domestic waste water is first treated in a conventional sewage treatment plant involving primary treatment followed by activated sludge. It is then retained for 30 days in an oxidation pond from which it is pumped to a spreading ground in coarse gravel which fills an old riverbed to a depth of 10 to 12 ft. After traveling through the gravel a distance of some 2,500 ft it is collected, chlorinated, and introduced into a series of ponds in which fishing and boating are permitted, and around which picnicking and other recreational activities are encouraged. During the summer of 1965 a special pond for swimming activities was operated safely and satisfactorily.

Restraints on the recreational use of the water were, of course, imposed pending the outcome of studies of travel of viruses.

Virologic tests reported [17, 18] in the period 1962 to 1964 give the following results:

(*a*) Samples of raw sewage, primary effluent, and activated sludge effluent were 100 percent positive. (Thirteen different viruses were identified.)

(*b*) Effluent from oxidation pond (30 days detention) showed 30 percent of samples positive.

(*c*) Recreational pond influent, after 2,500 ft in soil system, was 100 percent negative.

In 1964 a special study involving the introduction of attenuated polio virus in the water reaching the spreading ground was conducted [19]. Sampling wells were located at distances of 200, 400, and 1,500 ft down the wash. No virus was recovered at any of the sampling wells.

On the basis of these data it may be concluded that the soil mantle of the earth removes viruses, as it does bacteria. They would seem to justify the growing feeling among health authorities that reclaimed water should be passed through the soil before being mingled with the water resource from which water supply may be drawn.

4. *Rates of recharge*   The rates at which sewage effluents may be infiltered from surface spreading vary greatly with soil conditions, surface treatment, operational techniques, and other factors discussed in Chapter 16. Values of 0.5 to 1.0 acre-ft/acre/day are not uncommon values for fine soils. Direct injection rates of more than 7.5 gpm/ft depth of aquifer have seldom been observed [7]. In deep strata, values as low as 0.5 gpm/ft have been utilized for industrial wastes. Intermittent loading and resting of surface ponds is a necessary technique; and chlorination and redevelopment of recharge wells on a short-term

schedule are necessary [14] unless a highly clarified and chlorinated water is injected.

EXTENT OF PRACTICE

Groundwater recharge with sewage treatment-plant effluents has been practiced at many locations in Southern California to various degrees for many years. For the most part the objective of these undertakings has been disposal of effluent in a manner acceptable to water quality control authorities. Recently, however, attention has been turned to the long-advocated concept of deliberate purposeful reclamation of water from municipal return flows and its conservation by groundwater recharge.

The most extensive development of water reclamation from municipal wastes is the Whittier Narrows Project of the Los Angeles County Sanitation Districts [1]. Here the Sanitation Districts produce 10 mgd of reclaimed water. The Los Angeles County Flood Control District spreads this water on its existing spreading grounds. The Central and West Basin Water Replenishment District pays the Sanitation Districts for reclaimed water at the same rate charged by the Metropolitan Water District for Colorado River water used by the Replenishment District for groundwater recharge purposes.

This plant has proved to be an outstanding success and is currently being enlarged. At an early date it and other water reclamation plants of the district will be producing 65 mgd of reclaimed water.

The problem of ownership of recharged water has been resolved by the water replenishment districts of the area by charging a withdrawal fee to all who pump from the underground waters of the district. This fee is then used to purchase reclaimed or untreated Colorado River water to make up the overdraft on the groundwater basin.

Direct injection of reclaimed water into underground aquifers is a growing practice throughout the United States [8, 9]. However, since it bypasses the biologically active zone of the soil in which degradation or organic compounds takes place, water quality control agencies are increasingly disinclined to accept the method as suitable for waters of a quality which would not be accepted as undiluted groundwater. Consequently, direct injection is generally a method of conserving or managing water resources rather than one of water quality management.

In the context of quality management, injection may be used to dispose of waste waters not acceptable to the freshwater resource, and hence indirectly it becomes a factor in quality control. For example, oil and chemical companies have for some years been disposing of chemical and brine wastes by injection into deep strata below the groundwater. Certain radioactive wastes are likewise being so discharged. As a water reclamation or conservation technique, however, examples range from the California coastal areas where injected reclaimed water is holding back intruding saline water from the ocean, to Long Island, New York, where water withdrawn for cooling purposes is returned to the groundwater resource. Similar examples are to be found in many other states throughout the nation.

CONCLUSIONS

From the standpoint of water resources management it is evident that groundwater recharge with water reclaimed from municipal return flows is destined to increase in importance in the years ahead. Its role in water quality management is complex and hence difficult to assess. Here again, no simple equilibrium is involved. Direct reuse of reclaimed water by industry reduces withdrawals from the resource pool in favor of recycled water. However, since the increment of salinity added by use goes into the water supply, reused water is more saline than once-through water. The result is that water withdrawn for domestic use is returned to the atmosphere via industrial consumption rather than returned to the resource. However, since all the dissolved minerals return to the resource, if the resource was indeed a pool the net effect of reclamation on water quality would be zero. The same is true of groundwater recharge with used water which has not been desalted.

The value of water reclamation from a quality management viewpoint then rests upon its altering the interchange system so that the effect of beneficial use is transferred from one sector of the water resource to another where quality is less important. Planning for such a condition is far beyond the concepts of water reclamation thus far developed. The important fact, however, is that while reclamation of the sort herein discussed may alter the quality interchange system illustrated in Chapter 3, for either good or evil, it does not alter the buildup in salinity of the overall water resource which results from man's use of water. Eventually, desalination must become a major aspect of water quality control. Therein lies the importance of water reclamation in the future.

REFERENCES

1 J. D. Parkhurst: "Progress in Waste Water Re-use in Southern California," *J. Irrigation Drainage Div., Am. Soc. Civil Engrs.*, ASCE Proc. 4523, IR 1, March, 1965.

2 W. J. Oswald, C. G. Golueke, and H. K. Gee: "Waste Water Reclamation through Production of Algae," Contribution No. 22, Water Resources Center,

Sanitary Engineering Research Laboratory, University of California, Berkeley, August, 1959.

3 C. G. Golueke, W. J. Oswald, and H. K. Gee: "Harvesting and Processing Sewage-grown Planktonic Algae," SERL Rept. 64-8, Sanitary Engineering Research Laboratory, University of California, Berkeley, September, 1964.

4 W. J. Kaufman and R. L. Sanks: "Partial Demineralization of Brackish Waters by Ion Exchange," *J. Sanit. Eng. Div., Am. Soc. Civil Engrs.*, ASCE 92(SA6), December, 1966.

5 R. L. Sanks: "Partial Demineralization of Saline Sewage by Ion Exchange. Progress Report No. 1," University of California, Berkeley, October, 1964.

6 P. H. McGauhey and J. H. Winneberger: "Summary Report on Causes and Failure of Septic Tank Percolation Systems," Technical Studies Rept. FHA 533, Federal Housing Administration, Washington, D.C., April, 1964.

7 "Studies in Water Reclamation," Tech. Bull. 13, I.E.R. Series 37, Sanitary Engineering Research Laboratory, University of California, 1955.

8 R. F. Goudey: "Sewage Reclamation Plant for Los Angeles," *Western Construction News,* 5(20), October, 1930.

9 C. E. Arnold, H. E. Hedger, and A. M. Rawn: *Report upon Reclamation of Water from Sewage and Industrial Wastes in L.A. County, Calif.,* Planograph, April, 1949. (Abstracted in *Pub. Health Engr.* Abstr. xxx:5:23 (1950.)

10 "Studies of Waste Water Reclamation and Utilization," California State Water Quality Control Board Publication 9, 1954.

11 "Waste Water Reclamation in Relation to Ground Water Pollution," California State Water Quality Control Board, 1953.

12 Leonard Schiff: "Some Developments in Water Spreading," Provisional Report, Soil Conservation Service, U.S. Department of Agriculture, September, 1952. Supplemented November, 1953.

13 Leonard Schiff: "Water Spreading for Storage Underground," *Agr. Eng.,* 35(11), November, 1954.

14 "Investigation of Travel of Pollution," California State Water Quality Control Board Publication 11, 1954.

15 *Waste Water Reclamation at Whittier Narrows,* California State Water Quality Control Board Publication 33, 1965.

16 J. B. Askew et al.: "Microbiology of Reclaimed Water from Sewage for Recreational Use at Santee, Calif.," APHA, 1963.

17 J. B. Askew: "Virological Study Report for Santee Water Recreation Project," Report to California State Department of Public Health, January 28, 1965.

18 J. B. Askew: First to Ninth Quarterly Reports to California State Water Quality Control Board, 1962 to 1964.

19 *Microbiological Content of Domestic Waste Waters Used for Recreational Purposes,* California State Water Quality Control Board Publication 32, 1965.

CHAPTER 16

Quality management
by engineered soil systems

INTRODUCTION

The existence and periodic replenishment of groundwater as a geological and hydrological phenomenon demonstrate that it is possible for water to infilter the soil surface and to percolate downward through unsaturated soil or be translated laterally through aquifers. Furthermore, the relative absence of bacteria and suspended solids in spring and well water gives advance notice of important phenomena relative to pollution travel. There are several other phenomena related to the behavior of water in the soil mantle of the earth which might be surmised from gross observation or everyday experience.

1. Surface runoff during rainfall suggests that infiltration rates are subject to finite limits.
2. Failure of intense short-duration rains to moisten dry soils to any appreciable depth reveals the fact of air locking or gas binding.
3. A pond can be created only by applying water to the soil at a rate in excess of its infiltrative capacity, hence ponding on the surface or in subsurface systems bespeaks either a temporary or a permanent overloading of the soil.
4. It is possible, as demonstrated by the engineering science of soil mechanics, to rearrange soil particles and control soil moisture in such a way as to make the soil essentially watertight. Hence, inadvertent management of a soil system can result in loss of infiltrative or percolative capacity in accordance with the same laws.

Although these facts may seem obvious, man's record of comprehension of their implications does him little credit. For example, despite the fact that rain has been falling through a bacteria-laden atmosphere and passing through the biologically active zone of the earth for centuries, only to return to the surface bacteria-free, we have assumed that bacteria may travel freely with percolating water. And to protect ourselves from this assumption, states have passed laws prohibiting the introduction into the groundwater, in any systematic engineered way, of waste water of a quality unsuited for public supply. Haphazard ignorant discharges via septic tanks were not included in the concept. There are, however, several reasons why such laws were formulated and interpreted as they have been!

1. Lack of scientific knowledge of the behavior of bacteria in soils in relation to percolating water.
2. The need for large factors of safety (10 or more) where health considerations are involved and no precise knowledge exists.
3. The certainty that in fractured and fissured rock, in dissolution caverns and channels in limestone, and in very coarse materials, bacteria do indeed travel with moving water.
4. The problem of enforcement and administration of the law. Specifically: If the law is to differentiate between underground conditions where bacterial travel is and is not a hazard, a whole set of guides, parameters, and "thou-shalt-nots" has to be spelled out; a permit system must be established; and staff, budget, and procedure must be set up for managing the system.

The simplest, and perhaps wisest, course was to prohibit all groundwater recharge with unpotable water, at least until there should arise a compelling need to do otherwise.

NEED FOR ENGINEERED SOIL SYSTEMS

A need for man to exploit nature's groundwater replenishment phenomena through rationally designed systems derives from the many factors which make necessary his more intensive use of the available freshwater resource (see Chapter 4). And because much of the water which might be recharged to the earth is return flow from beneficial uses, it is particularly important that the ability of the soil to change the quality of water be fully understood and systematically utilized as a quality management technique.

The specific need for engineered soil systems may be outlined as follows:

1. Salvage and conservation of
   (a) Floodwaters
   (b) Reclaimed water
2. Utilization of underground storage capacity
3. Providing waste-water disposal for
   (a) Rural homes and institutions
   (b) Urban and suburban subdivisions
   (c) Isolated commercial enterprises

4. Utilization of biologically active soil mantle of the earth as a waste-water treatment system for
   (a) Protecting groundwater quality
   (b) Reclaiming water from waste-water discharges

TYPES OF SOIL SYSTEMS

The types of engineered systems which may be utilized for ground disposal of waste waters include:

1. Surface infiltration ponds (spreading ponds) or trenches
2. Injection wells
3. Subsurface percolation systems
   (a) Narrow trench
   (b) Wide trench—seepage bed
   (c) Seepage pit
   (d) Cesspool
4. Sand or soil filters (effluent returned to surface)

Of these several types the infiltration pond is the most used for ground-water recharge operations; the injection well, for deep strata waste discharge; the subsurface system, for septic-tank effluents; and the sand filter, for water reclamation. All, however, involve the same fundamental principles.

TYPES OF ENGINEERING
PROBLEMS INVOLVED

In relation to the utilization of the foregoing types of soil systems, two types of engineering problems have been of particular concern to engineers:

1. How to get water into the soil at a maximum rate
2. How to achieve optimum quality change in water passing through the soil system

The first of these two objectives has been given the most attention for obvious reasons. First, man cannot afford to sprinkle water over a large area intermittently as does nature. Furthermore, he is greedy. Consequently he has sought to introduce water into the earth continuously and at rates vastly greater than occur in nature. The dimension of his failure has called attention to the simple lessons noted in the introductory paragraph of this chapter.

Attention to the second type of problem began with studies of the travel of pollution with percolating water [1] and with groundwater [2], as a result of constraints placed upon groundwater recharge with sewage effluents by public health legislation. The objective was to determine what types of quality factors move with underground waters and under what conditions.

Currently the inverse of this objective is under study [3, 4, 5], i.e., the management of soil systems to produce an optimum change in water quality. The approach is from two important viewpoints:

1. In the case of a specific quality requirement for the groundwater resource, what quality of water (hence what degree of pretreatment) may be applied to the soil surface without endangering groundwater quality?
2. In the case of a recharge water of any observed quality, what will be the quality of that same water upon reaching the groundwater table?

Practical solutions to both types of engineering problems noted depend upon application of fundamentals such as those hereinafter summarized.

FUNDAMENTAL ASPECTS OF SOIL SYSTEMS [3, 4, 6]

It is obvious to even the most casual observer that in order to create a pond it is necessary to apply water to soil faster than the soil can accept it and transport it away. The phenomena involved, however, are far from obvious, yet the whole problem of disposing of water successfully by use of spreading ponds or percolation systems hinges upon a knowledge of these phenomena and the methods of operation required to manage them in a particular situation.

Of course, if water is to be disposed of by percolation it must be applied to a soil that is pervious enough to carry it away at an acceptable rate. It is well known that structureless clays are quite watertight and hence unsuitable for recharge systems, whereas coarse gravel may accept water at such a rate that ponding of applied water may not occur. Curiously enough, the porosity of the two media may be quite similar; it is their perviousness that is different. For practical purposes these two terms may be defined as follows:

Porosity is the percentage of a volume of material that is void space.

Perviousness refers to the size of the void spaces involved in porosity. It is, however, measured by the rate of passage of liquid under standard conditions. This measure is called the *permeability* of the system.

Perviousness of a soil, and its resultant permeability, to which water is to be disposed, is important for at least two major reasons:

1. It controls the maximum rate at which water can percolate downward to the water table or be translated laterally as moving groundwater.
2. It controls the minimum elevation of the infiltrative surface above the water table in any soil having pore spaces small enough to establish surface tension of water across the pore as a significant force. Specifically, the water table must be located far enough below the infiltra-

tive surface to permit the soil itself to drain when no more water is applied to the infiltrative surface. If such is not the case, a "hanging column" of water will remain in the soil pores, suspended by the surface tension of water across the pores. This, as explained later, is ruinous to a percolative system. In good agricultural soils observed in California the critical distance is of the order of 2 to 2.5 ft. In coarse gravel it is zero, and in clays it may theoretically be infinity.

Two additional concepts underlie the practical functioning of a soil percolation system.

1. The *infiltrative capacity,* or rate at which liquid will pass through the soil-water interface. It measures the ability of a soil to accept water.
2. The *percolative capacity,* or rate at which water moves through the soil once it has passed the interface. It measures the ability of a soil to transport water.

TIME RATE OF INFILTRATION

One the most fundamental of all phenomena of surface application of water to soil is the well-known time-rate infiltrative curve. If a clear water without suspended solids or dissolved organic matter is applied continuously to a pervious soil, the time-rate curve will appear somewhat as shown in Figure 16-1.

If the effluent from a conventional sewage treatment plant or a septic tank is used instead of clear water, the curve shown in Figure 16-1 is modified in shape. Organic nutrients in the sewage effluent are so vast in comparison

Figure 16-1 *Typical time-rate infiltration curve for water.*

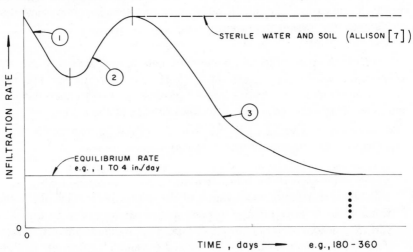

with those normally found in soil that a rapid growth of organisms overwhelms both sectors **1** and **2**. Consequently, the time-rate infiltration curve for sewage plunges steeply downward from the very beginning, then approaches the same equilibrium conditions shown in Figure 16-1, albeit at an earlier time because of the greater clogging potential of sewage. In reality an equilibrium is never completely reached, but the decline eventually becomes quite slow.

The phenomena predominant in the three sectors of the infiltrative curve have been identified as follows:

1 Soil slaked due to

  (a) affinity of internal soil surface for water

  (b) overcoming cohesive forces which hold soil together

2 Entrapped air removed by solution in percolating water.

3 Decline in infiltration rate due primarily to microbial action in the soil, particularly to the rise of anaerobic organisms feeding on organic matter in a soil from which air has been displaced by water. (Allison [7] noted that decline does not occur with sterile water in a sterile soil.)

The high point of sector **3** of the curve represents, to an acceptable degree, the percolative capacity of the soil. It also represents the initial infiltrative capacity of the soil system. Infiltrative capacity, however, declines along sector **3** due to soil clogging, whereas the percolative capacity of the soil remains constant.

Evaluation of the percolative capacity of a soil in its natural bed is customarily attempted by a "standard percolation test" [8], the results of which are then interpreted in terms of infiltrative capacity on the basis of curves or tables originally derived from limited field observation [9]. The test is subject to many serious limitations [3, 10], both theoretical and practical. Perhaps the greatest fallacy in past efforts to interpret the test in septic-tank percolation-field design, where it is most commonly used, is the assumption that by observing the short-term percolative capacity of a soil it is possible to predict the permeability of an organic clogging mat which later develops at the soil surface.

Several important facts may be deduced from a consideration of the infiltrative-percolative-capacity relationships of a homogeneous soil:

1. The percolative capacity of a soil may be estimated by a percolation test if properly conducted at a sufficient number of locations on the soil area to be inundated.

2. The percolation test can at best only identify a soil capable of transporting water (and hence suitable for spreading purposes) at an observed rate, provided infiltrative capacity can be maintained.

3. The infiltrative capacity of a soil, except when water is first applied, is always less than the percolative capacity by reason of soil clogging which, as later explained, is largely a surface phenomenon.

4. The infiltrative capacity is variable and is a function of the numerous factors which are involved in soil clogging. Hence it cannot be predicted in advance by currently known tests.

5. Since the infiltrative capacity declines below the percolative capacity along a die-away curve, the problem of design and operation of an engineered soil system is that of *maintaining the infiltrative capacity of the system as near as possible equal to its percolative capacity.*

OPTIMIZING
INFILTRATION RATES

Inasmuch as the initial high rate of infiltration (percolative capacity) cannot be maintained, the equilibrium infiltration rate becomes the controlling factor. It has been repeatedly shown experimentally that this equilibrium rate is independent of the initial rate and reaches a quite constant value regardless of the nature of the soil itself. This is because it is governed by the nature of the organic clogging mat rather than by that of the soil which supports it. In fact, a coarse material behaves no better than a fine soil once it is clogged. The important question, then, is whether or not the equilibrium infiltration rate is great enough to be of practical use in an engineered system. Data such as shown in Table 16-1 indicate that the equilibrium rate for continuously inundated soils becomes negligible within periods of time quite short in comparison with the useful life expected of an engineered system.

The data likewise show clearly (column 5 versus column 6) the lack of relationship between percolative capacity and equilibrium infiltrative capacity. They show also that in general the coarser materials with small coefficients of uniformity have high percolative capacities.

Experiments on the effect of cyclical loading show that in contrast with the typical time-rate curve of Figure 16-1, periods of rest and loading will reproduce the upper section of sector **3** in a pattern such as the following (Figure 16-2), thus permitting long-time use of an infiltrative surface at acceptable infiltration rates.

The optimum cycle of resting and loading must be determined for each soil situation. Orlob and Butler [11], however, demonstrated the feasibility of predicting the behavior of undisturbed homogeneous soils in their natural beds by use of small (30-in. diameter) lysimeters packed with 3 ft of disturbed soil. Typical curves for each soil, such as illustrated in Figure 16-1, were obtained. By alternate periods of loading and resting, a curve similar to Figure 16-2 was obtained for each soil, yielding a basis for estimating the long-term sustainable infiltrative capacity of the system.

Operation of spreading basins for optimizing the rate of infiltration has been studied for most of the soils listed in Table 16-1. The optimum period of loading and resting of Hanford soil when sewage effluents are used is approximately 2 weeks of loading, followed by 1 week of resting. In Israel the period is about the same but in months instead of weeks. Restora-

Figure 16-2 *Typical restoration of infiltrative capacity by cyclical operation.*

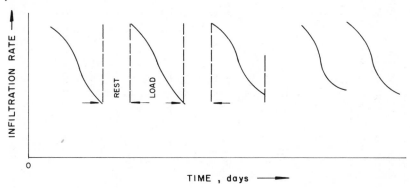

tion of infiltration rate observed [11] in the Hanford soil was from seven to ten times the equilibrium rate. For 163 days the gain was ten times equilibrium, and the net for 28 months was five times. In the winter, rate restoration was smaller than in the summer.

At the present state of technology, the optimum cycle of loading and resting must be determined by experiment with the finished installation.

The curves shown in Figure 16-2 are typical of horizontal surfaces in spreading ponds where both draining and drying of the soil may occur during periods of resting. A much more complex situation develops in infiltration trenches or septic-tank percolation trenches or pits where vertical sidewalls represent the infiltrative surface. For instance, if the system, e.g., trench or pit, were completely filled at the start of operation, each successive soil horizon would be subject to a longer period of inundation than the one above it. Thus the topmost horizon might have a very short period of loading and a long period of rest, whereas the bottom horizon might get no rest at all. The result would be that a recovery curve at the top of the wall might look like Figure 16-2, and at the bottom of the wall, like the extreme right-hand sector of the die-away curve in Figure 16-1. The restoration pattern for the sidewall surface would then be a composite of these end curves and all intermediate curves, unless the time of rest required to restore the infiltrative capacity of the very lowest horizon were provided.

The effect would be a downward shift of the trend of the restoration

TABLE 16-1 *Equilibrium rates of infiltration into various soils*

Soil	General nature	Effective size, mm	Uniform coefficient	Initial infiltration rate, ft/day	Equilibrium infiltration rate,* ft/day	Depth of submergence, ft	Source of data or reference no.
(Azusa plot)	Coarse alluvium	0.40	68	...	1.2	1.0	
Hesperia	Sandy loam	0.002	67.3	5.2	0.5	1.0	
Hanford	Fine sandy loam	0.0074	24.9	6.4	0.3	1.0	[11]
Yolo	Sandy loam	0.021	8.1	10.0	0.3	1.0	
Oakley	Fine sand	0.020	11.2	29.0	0.16	1.0	
Columbia	Sandy loam	0.0033	47.3	1.8	0.13	1.0	
(Whittier plot)	Sandy loam	0.044	4.3	...	0.6	0.5	
Hanford (Lodi)	Fine sandy loam	0.0032	67	1.1	0.58	0.5	[11]
		0.0035	86	1.1	0.25	0.5	
		0.0032	67	1.2	0.58	0.5	
		0.0018	78	1.1	0.17	0.5	
(Israel)	Dune sand: Lysimeters	0.16	1.44	15	0.33	$\simeq 2$	[12]
	Natural state	0.16	1.44	16	0.67	$\simeq 3$	
(Gonzales)	Alluvial sand	...	...	...	0.10	5	Field observations†
		...	...	...	$\simeq 0.05$†	5	Estimate‡
(Soledad ponds)	Alluvial sand	$\simeq 0.15$	$\simeq 2.4$	...	< 0.065	±5	Calculated from field data
Hole No. 3	Alluvial sand	0.14	2.1	...	...	...	
Hole No. 4		0.16	2.6	...	...	...	

*Approximate steady state after continuous flooding with primary effluent for periods ranging from a few weeks to 2½ years.

†W. J. Oswald.   ‡After 18 months raw sewage ponding. Estimate by Soledad City Engineer.

curves until creeping failure, as later described, had reduced the infiltration rate of the overall system to an unacceptably low value. At this time a very long period of rest (probably months) would be required for the infiltrative surface to recover entirely.

The foregoing factors would lead theoretically to a curve pattern such as shown in Figure 16-3 in which successive curves are lower in elevation and steeper in descent.

Figure 16-3 *Theoretical pattern of restoration of infiltrative capacity of a trench sidewall.*

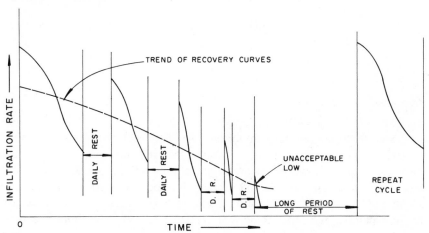

Here again the optimum cycle of resting and loading must be determined in each situation.

CLOGGING OF SOILS

A soil of such characteristics that it has a positive percolative capacity may become clogged as a result of physical, chemical, and biological factors, although ultimately the exclusion of water becomes a physical phenomenon.

Physical factors in clogging

There are several ways in which clogging may result from predominantly physical phenomena. They include:

1. Compaction of soil by superimposed loads (e.g., ponded water, heavy equipment, etc.)
2. Smearing of soil surfaces by excavating equipment
3. Migration of fines by vibration of dry soil during preparation of site
4. Migration of fines due to rainfall beating against surface

5.   Washdown of fines perched on larger particles

The travel of particles in a porous medium is inhibited by at least three major factors which lead to clogging by particulate matter. These are illustrated in Figure 16-4.

Figure 16-4   *Removal of particles from water through a soil system.*

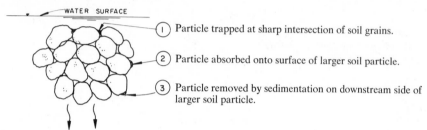

Particle trapped at sharp intersection of soil grains.

Particle absorbed onto surface of larger soil particle.

Particle removed by sedimentation on downstream side of larger soil particle.

Chemical factors in clogging

Of the purely chemical, as opposed to biochemical, phenomena, ion exchange is by far the most important. The deflocculation of clay colloids by sodium is a well-known way in which either soil surface or aquifers may become clogged. However, in areas where soils suffer from this limitation the fact is soon known and soil systems for sodium-laden waters are not attempted. Assuming, therefore, that soil, recharge water, and groundwater are compatible, the most important factor in clogging becomes the biological one.

Biological factors in clogging

Clogging by biological factors is essentially a surface phenomenon involving the development of an organic mat. In soils with modal size up to approximately 1 mm in diameter, this organic mat seldom exceeds 0.5 to 1 cm in depth. Primarily this mat consists of organic solids taken out by the three factors illustrated in Figure 16-4, plus an overgrowth of bacteria which feed on particulate and dissolved organic matter. As the porosity of the mat decreases, it becomes itself a filtering medium catching smaller and smaller solids and so decreasing the infiltrative capacity of the soil. It is alleviated by intermittent resting of infiltrative surface during which time:

1.   The surface is drained and oxygen drawn in to maintain high-rate aerobic decomposition;

2.   The clogging material is degraded biochemically to liquid and gases; and

3.   The drying and cracking may improve soil texture and increase its infiltrative capacity.

Overcoming the effects of biological clogging is thus the principal explanation of the restoration of infiltration rates illustrated in Figure 16-2. The action of draining and resting a soil is to keep the system itself aerobic. Both filling soil pores with oxygen and restoring an aerobic biota in equilibrium with its environment are involved in recovery. In the absence of a loading pattern which maintains aerobic conditions, anaerobic conditions set in. Excessive clogging then develops from two major factors as illustrated in Figure 16-5:

1. Rapid growth of slimes in an organic mat
2. Deposition of ferrous sulfide

Figure 16-5 *Elements of clogging zone in organic loaded soil.*

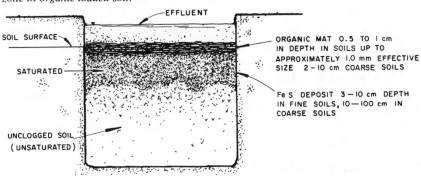

The organic mat illustrated in Figure 16-5 is by far the most significant phenomenon in the problem of soil system operation, particularly where domestic waste waters are applied. Within a quite wide range of particle sizes, the pervious soil serves largely as a matrix to support a mat of relatively constant permeability little related to that of the supporting material. This mat controls the rate of passage of liquid through the system, and hence the apparent equilibrium rate of infiltration of sewage effluent is not widely different from one soil to another, except in soils so fine that percolative capacity controls, or so coarse that the organic mat is dispersed in depth rather than concentrated at the soil surface. Unfortunately, the characteristics of the organic mat as a material under aerobic and anaerobic conditions have not been investigated, research having only recently isolated it as a phenomenon separate from soil characteristics.

Ferrous sulfide is a black particulate matter that gives the characteristic color to digested sewage sludge, anaerobic compost, clogged septic-tank percolation systems, etc. It is the result of anaerobic decomposition of organic matter and its presence in appreciable amounts indicates an unsatisfactory condition of operation of the percolation system intended to dispose of sewage effluent to the soil. It may account for a large percentage of the loss of infiltrative capacity.

In contrast with the biological clogging mat, which may be only ½ to 1 cm in depth on a fine pervious soil, ferrous sulfide may penetrate 3 to 10 cm below the surface just as may any other finely divided inert material. In coarser media it may penetrate to much greater depths, as may the bacteria which produce it. Deep penetration of ferrous sulfide in a soil is indicative that the soil has a good capacity to accept water if properly managed.

Fortunately ferrous sulfide is readily oxidized to the soluble sulfate form upon resting and draining of the soil to bring in atmospheric oxygen. However, a pervious soil that becomes deeply clogged with ferrous sulfide to the extent that it remains waterlogged may remain clogged indefinitely. Drainage to fill the pore spaces with air is a necessary factor in removal of ferrous sulfide.

Interestingly enough, the formation of ferrous sulfide cannot be prevented by simply blowing air through a septic sewage; the medium itself must be drained and rested.

The appropriate criteria of soil systems operation and design for maximum rates of infiltration then are: Maintain an aerobic system by alternate periods of resting and loading on an optimum cycle; and, reduce to a minimum by pretreatment the biochemically unstable organic content (suspended and dissolved) of the applied water.

Clogging during construction

One of the major causes of failure of percolation systems is the damage done by the careless and ignorant methods used in construction. Among the most serious causes of permanent loss of infiltrative capacity are:

1. Smearing of sidewall and bottom surfaces during construction
2. Compaction of bottom surface by human feet or by dozer tracks or wheels
3. Silting of open excavation during rain, by spalling of walls, or by windblown loess

Damage by smearing and compaction is most severe where soil particles are small and soil moisture is high. *The criterion here must be that of providing an initial infiltrative surface as near as possible representative of an internal plane in the undisturbed soil.* It seems likely that construction methods adequate to this goal might be delicate and costly as compared to current procedures.

Creeping failure

The geometry of septic-tank percolation systems, together with a lack of understanding of infiltrative phenomena, leads to continuous ponding and eventual failure in a manner which the author has termed [3] "creeping failure." Creeping failure is best illustrated by the small-diameter seepage pit depicted in Figure 16-6.

In the seepage pit the discharge of effluent into the pit at time t results in a shallow pond. Thus only a small increment of wall surface is loaded. Only as this ring is overloaded and heavily clogged does a ring at the next

Figure 16-6 *Failure of infiltrative surface by incremental overloading.*

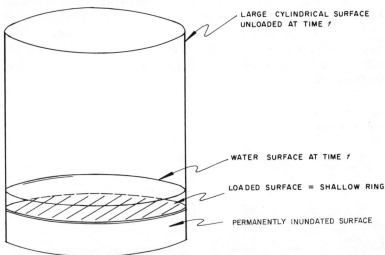

LARGE CYLINDRICAL SURFACE UNLOADED AT TIME t

WATER SURFACE AT TIME t

LOADED SURFACE = SHALLOW RING

PERMANENTLY INUNDATED SURFACE

higher elevation receive liquid. Hence increment by increment the wall is loaded to failure, the unloaded infiltrative wall surface being totally useless until the rising flood reaches it.

The same situation is true of the narrow trench. Here the bottom is soon clogged and both longitudinally and vertically the sidewall is overloaded. In the wide trench the bottom first functions as an infiltrative surface, hence such a system is analogous to a surface spreading pond rather than a pit or narrow trench. Creeping failure in this case first takes the form of advancing contours from the point of introduction of effluent to the system [3]. In practice, however, the bottom eventually clogs and inadequate sidewall areas then begin to fail in the manner of the trench or pit.

Thus the appropriate criterion of design and operation is: *The entire infiltrative surface should be loaded uniformly and simultaneously,* an ideal which can be achieved in spreading ponds but not in trenches or in septic-tank percolation systems. Application of the criterion, however, does not prevent creeping failure from becoming the major clogging routine in any system with vertical sidewall infiltration surfaces. The phenomenon depicted in Figure 16-3 is unavoidable and must figure in any theoretical or practical analysis of a soil system having vertical infiltrative surfaces, primarily because time for full restoration of lower horizons is prohibitively long for application to each cycle of loading and resting.

Effect of trench fill stone

An important loss of infiltrative surface in underground systems, and hence of infiltrative capacity, results from a sealing of the soil surface if there is a great disparity in particle size between gravel fill and soil particles in the leaching field. Figure 16-7 illustrates the phenomenon of surface sealing for either a sidewall or a bottom condition. Here an important fraction of the infiltrative surface is lost due to the "shadow zone."

Figure 16-7   *Loss of infiltrative capacity by abrupt change in particle size.*

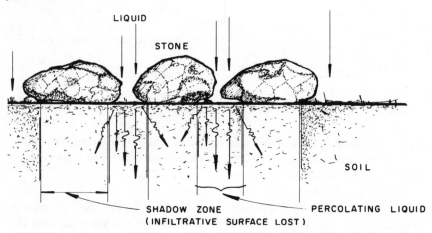

LIQUID

STONE

SOIL

SHADOW   ZONE   PERCOLATING   LIQUID
(INFILTRATIVE   SURFACE   LOST)

Surface sealing, however, is not the only clogging effect resulting from the use of relatively large-size trench fill stone. Intrusion of soil into the voids in the stone fill has been observed. Such intruded material loses the structure which characterizes the soil in its natural bed, the fines may wash down, and the whole mass becomes mixed with biological slimes and flows over the infiltrative surface, thus effectively sealing it. In addition, large stones have a minimum of the inhibiting effect on passage of suspended matter (illustrated in Figure 16-4). The criterion is that *there should be no abrupt change in particle size between trench fill material and infiltrative surface of the soil.*

Clogging in depth

The often observed high infiltrative capacity of coarse soils and sand over a prolonged period is not strictly proportional to the size of openings which govern the permeability of the material. As previously noted, for soils up to about 1 mm in modal size the nature of the organic mat is unrelated to the permeability of the underlying soil. Beyond some critical pore size, however, suspended matter and organic nutrients, and the bacteria which feed upon them, penetrate the soil and are removed by the forces depicted in Figure 16-4 in a depth of 2 or 3 cm or more, rather than at the surface. Thus the organic mat is distributed in depth and has a less profound clogging effect than if it were concentrated at the surface.

Fine soils which have had their surface structure altered by agglomeration or have been covered with organic trash and similar coarse material have likewise, after a period of incubation, been shown experimentally [13] to clog in depth and to exhibit an infiltrative capacity greater than normal. In controlled studies [14] it has been shown, by overlaying a soil with material graded from the soil particle size at the bottom to fine gravel at the top, that *clogging* in depth is an important factor in increasing and sustaining infiltrative capacity.

SIDEWALL VERSUS BOTTOM SURFACES

Except in shallow ponds where infiltration of the soil depends upon the ability of the pond bottom to accept water, the sidewalls of a percolation system are by far the most effective infiltrative surface. Until experiments demonstrated this truth, it was largely overlooked except in the seepage pit. In fact, the whole practice of narrow trench disposal of septic-tank effluents, complete with design tables, was built up and codified, apparently without discovering that its principal criterion—the area of trench bottom—was irrelevant and meaningless. The now established effectiveness of sidewall areas in trenches and pits derives from several factors, of which the following are the most important:

1. The process of sedimentation is not a major factor in sidewall clogging.
2. Declining water depth in a percolation trench provides alternate resting and loading of the sidewall surfaces (although not all increments are subject to the same periodicity), whereas the bottom may be continuously inundated.
3. Gravity assists in stripping sidewalls of clogging material as it decomposes during periods of resting.

Since all factors associated with loss of infiltrative capacity are most serious on the bottom surface, the criterion of soil system design becomes:

Trench systems should provide a maximum of sidewall surface per unit volume of effluent and a minimum of bottom surface.

ENGINEERING OF SOIL SYSTEMS FOR MAXIMUM INFILTRATION RATES

As is the case with other engineered systems for water quality management, there are no well-developed methods of proceeding from theory to design criteria, although numerous attempts have been made to relate sustainable infiltration rates to some characteristic of the soil. Nevertheless, the principles and procedures for the engineering of soil systems for maximum infiltration rates are now well, if not widely, understood. Some of the most important are summarized in Table 16-2.

ENGINEERING OF SOIL SYSTEMS FOR OPTIMUM QUALITY CHANGE

As previously noted, there is little experience in engineering of soil systems for the specific purpose of changing the quality of water. A recognition of the ability of soil to act as a treatment system, however, has resulted from both a rational analysis of natural phenomena and from extensive research. In Chapter 15 evidence is cited of the removal of bacteria and viruses by soils. The mat of organic matter which clogs a soil surface represents a biological system analogous to the trickling filter or activated sludge processes, and its decline during periods of resting is largely the result of aerobic degradation and stabilization of organic matter which formed the mat. Thus soluble products of biodegradation may be expected to travel downward with percolating water once any capacity of the soil to adsorb them has been satisfied. With increasing particle size, bacteria and organic matter penetrate deeper into a soil and hence delay the progress of clogging. Beyond some critical size the removal of particulate matter ceases to be fully effective and bacteria themselves move through the soil. At that critical size, however, it is conceivable that the rate of loading can exceed the capacity of the system to degrade organic matter and so allow partially stabilized products or dissolved organic solids to escape beyond the biologically active zone. These may impart tastes and odors to groundwater or support bacterial life in a well. Effective purification of sewage effluent by bacteria present several feet below the soil surface, probably in a surface buried during construction of the spreading basin, has been observed [15] in California.

Several experiments [16, 17, 18] have demonstrated the ability of soil systems to degrade such synthetic organic compounds as ABS, LAS, and other surfactants. In such cases the effectiveness depends upon the development of a microflora adapted to the compounds present. Thus the optimum

TABLE 16-2 *Summary of principal factors in engineering of soil systems for maximum infiltration rates*

Objective or criterion	Method of achievement
Surface spreading or infiltration basins	
Evaluation of soil	Use Soil Conservation Service data to identify local soils which will accept rainwater and drain. Make geological study and soil borings on proposed site to determine presence of limiting zones (clay lenses, hardpan, bedrock, etc.). Make replicate percolation tests to estimate maximum percolative capacity (soil permeability). Determine by laboratory test the compatibility of soil and water to be applied (ion exchange, dispersion of colloids).
Maintenance of aerobic system	Estimate by lysimiter tests or pilot spreading ponds the optimum cycle of loading and resting. Design for multiple basins with factor of safety to allow for refinement of cycle on basis of operating experience. Provide factor of safety in excess land area.
Minimizing unstable organic content of applied water; minimizing of suspended particles	Provide feasible maximum of pretreatment by conventional systems, oxidation ponds, chlorination, etc.
Provision of undamaged initial infiltrative surface (initial infiltrative capacity equal to percolative capacity of soil)	Prepare site by unconventional methods, and at time of most favorable soil moisture conditions, to avoid smearing, compaction, or other damage to infiltrative surface.
Loading infiltrative surface uniformly and simultaneously	Level bottom of spreading basin. Design basin size and inlet structures for rapid inundation of bottom surface (i.e., hours instead of days).

TABLE 16-2   *Summary of principal factors in engineering of soil systems for maximum infiltration rates (continued)*

Objective or criterion	Method of achievement
Surface spreading or infiltration basins (continued)	
Dispersing clogging mat in depth; avoiding abrupt change in particle size over soil surface	Overlay soil surface with coarser material in two or three layers of sand graded from fine at the bottom to coarse at the top. If sand overlay not used, treat surface in one of various ways, e.g., vegetation, organic trash, soil conditioner, scarifying, etc. (see literature). In any event, provide alternate periods of wetting and drying to agglomerate soil.
Providing maximum sidewall - to - bottom - surface ratio	(Not applicable to spreading ponds.)
Trenches and pits	
Evaluation of soil	(Same as for spreading basins.)
Maintenance of aerobic system	Design for multiple units. Determine optimum cycle of loading and resting by operating experience, after installation. Anticipate need for prolonged period of resting when creeping failure reduces infiltrative capacity of trench to minimum acceptable limits. Provide factor of safety in excess land area.
Minimizing unstable organics and suspended particles	Provide and maintain most efficient pretreatment system feasible.
Provision of undamaged infiltration surface	Use careful construction procedures (same as spreading basins).
Loading infiltrative surface uniformly and simultaneously	Use level bottom trenches. Provide multiple units and dosing system for quick filling of trenches or pits. (Not entirely feasible for septic tanks.) Use narrowest possible trench. Locate influent tile above useful trench wall.

TABLE 16-2 *Summary of principal
factors in engineering of soil systems
for maximum infiltration rates (continued)*

Objective or criterion	Method of achievement
Trenches and pits (continued)	
Dispersing clogging mat in depth; avoiding abrupt change in particle size	Fill trench with material graded from fine at trench wall to coarse at center. (Two sizes of material may be practical limit in narrow trench.)
Providing maximum sidewall - to - bottom - surface ratio	Utilize deep narrow trenches.
Subsurface seepage beds (spreading basins)	
(All objectives)	(Treat as combination spreading basin and trench. Suitable only for small-scale installations of doubtful utility.)
Injection wells	
Evaluation of aquifer	Examine well logs for presence of dispersible clay. Determine by laboratory tests the compatibility of aquifer material, injection water, and groundwater.
Determination of optimum injection rate	Estimate on basis of permeability of aquifer, depth of strata, thickness of overburden, and permissible wellhead pressure. Refine estimate ón basis of test well.
Prevention of clogging of aquifer	Provide maximum of pretreatment to minimize suspended and unstable matter in injection water. Chlorinate injection water. Use gravel-pack injection well. Provide for redevelopment of well.

loading and resting cycle for maximum infiltration rates is not necessarily the optimum for changing water quality, at least until microflora have been developed.

From an engineering viewpoint a natural soil system, or one especially set up for the purpose, can be considered as a matrix on which to hang a biological system.

A great deal of experience is already available with slow sand filtration of water supplies. Such artificially constructed soil systems have also been used with septic tanks in areas where the groundwater is near the surface, and in engineered water reclamation plants at Whittier Narrows and the Hyperion Sewage Treatment in Southern California. Parameters for design of such systems as filters are well known. Similar parameters for optimum biodegradation of organic compounds by such systems are capable of development. Although for reasons of economy an engineered soil system might normally involve the use of soil in its natural bed, there is no reason why soil particles should not be separated and reconstituted as a more suitable matrix for a biological system designed to effect predictable changes in waste-water quality.

As a method for quality management, direct injection into underground waters is not feasible. The system cannot be maintained in an aerobic condition because of saturation. A clogging mat will rapidly build up within and beyond the gravel pack if the recharge water contains particulate matter. If the particulates are unstable organic matter, an aerobic activity will produce only unstable breakdown products. These, together with dissolved organic matter also only partially degraded, will be carried by water moving at high velocity and pressure into the aquifer beyond the zone of biological activity. Consequently, direct injection must be viewed merely as a device for putting highly treated water into the earth rather than as a system for water renovation.

REFERENCES

1 *Field Investigation of Waste Water Reclamation in Relation to Ground Water Pollution*, California State Water Quality Control Board Publication 6, 1953.

2 *Investigation of Travel of Pollution*, California State Water Quality Control Board Publication 11, 1954.

3 P. H. McGauhey and J. H. Winneberger: "Causes and Prevention of Failure of Septic-tank Percolation Systems," Tech. Studies Rept. 533, Federal Housing Administration, April, 1964.

4 P. H. McGauhey and J. H. Winneberger: "Final Report on a Study of Methods of Preventing Failure of Septic-tank Percolation Systems," University of California SERL Rept. 65-17, October 31, 1965.

5 P. H. McGauhey, R. B. Krone, and J. H. Winneberger, "Soil Mantle as a Wastewater Treatment System," University of California SERL Rept. 66-7, September, 1966.

6 P. H. McGauhey: "Report on Technical Feasibility of Land Disposal of Sewage Effluent at the City of Soledad, California," Engineering Report to Cen-

tral Coastal Regional Water Quality Control Board, San Luis Obispo, California, June 14, 1965.

7 L. E. Allison: "Effect of Microorganisms on Permeability of Soil under Prolonged Submergence," *Soil Science*, 63, 1947.

8 Anon: *Manual of Septic-tank Practice*, U.S. Public Health Service Publication 526, 1958.

9 Henry Ryon: *Notes on the Design of Sewage Disposal Works with Special Reference to Small Installations*, Albany, N.Y.: Private publication, 1928.

10 P. H. McGauhey, G. T. Orlob, and J. H. Winneberger: "A Study of the Biological Aspects of Failure of Septic-tank Percolation Systems," First Progress Report, Report to Federal Housing Administration, Sanitary Engineering Research Laboratory, University of California, Berkeley, December, 1958.

11 J. T. Orlob and R. G. Butler: *An Investigation of Sewage Spreading on Five California Soils*, Technical Bulletin 12, I.E.R. ser. 37, no. 12, Sanitary Engineering Research Laboratory, University of California, Berkeley, June, 1955.

12 A. Amramy et al.: "Dan Region Sewage Reclamation Project-Infiltration and Percolation Studies," Progress Report, P.N. 247, TAHAL (Tel-Aviv), Water Planning for Israel Ltd., September, 1962.

13 L. Schiff: "Some Developments in Water Spreading," Provisional Report, Soil Conservation Service, U.S. Department of Agriculture, September, 1952. Supplemented November, 1953.

14 J. H. Winneberger, A. B. Menar, and P. H. McGauhey: "A Study of Methods of Preventing Failure of Septic-tank Percolation Fields—Second Annual Report," Report to Federal Housing Administration. Sanitary Engineering Research Laboratory, University of California, Berkeley, December 22, 1962.

15 *Waste Water Reclamation at Whittier Narrows*, California State Water Quality Control Board Publication 33, 1966.

16 S. A. Klein, D. Jenkins, and P. H. McGauhey: "The Fate of ABS in Soils and Plants," *J. Water Pollution Control Federation*, 35(5), May, 1963.

17 S. A. Klein and P. H. McGauhey: "Degradation of Biologically Soft Detergents of Wastewater Treatment Processes," *J. Water Pollution Control Federation*, 37(6), June, 1965.

18 G. G. Robeck et al.: Factors Influencing the Design and Operation of Soil Systems for Waste Treatment," Paper presented at Water Pollution Control Federation, Seattle, October, 1963.

CHAPTER 17

Quality management by engineered ponds*

INTRODUCTION

The purposeful addition of organic wastes to surface ponds antedates history. During ancient times in the Orient and in Europe, and at present in many places throughout the world, ponds fertilized with organic wastes, as well as with inorganic fertilizer, have been constructed and operated to encourage algal growth and thereby greatly increase the areal yield of fish or other aquatic organisms which feed directly or indirectly upon algae. The purification of sewage in fish ponds has been a recognized art in Germany for half a century. In America, however, fish ponds have not been used intentionally for sewage treatment. In fact the first waste stabilization ponds in the United States evidently were built for the sole purpose of excluding waste waters from other sectors of the water resource in which they would be objectionable. Once built, however, the waste purification potential of ponds became increasingly evident to the observer. Following a description of the treatment potential of certain accidentally formed ponds at Santa Rosa, California, in 1944 by Gillespie [1], a succession of papers described the behavior of several specific pond installations. More recently a growing body of technical and scientific literature [2, 3] has placed stabilization pond design on an increasingly rational basis.

It is now widely recognized that when properly designed and operated, stabilization ponds will develop a population of organisms which will degrade organic matter and subsequently convert the low-energy products of degradation into high-energy algal cells. If these cells are then removed from the liquid, the final effluent is of a quality superior to that produced by more conventional waste-treatment processes. In terms of the aerobic cycle of growth and decay presented in Chapter 2, it may be said that whereas the normal treatment system degrades organic matter to stable

*This chapter authored by W. J. Oswald.

210

fertilizer compounds which remain in the water without exerting an oxygen demand, the pond system goes further to pick up a large fraction of these fertilizers and incorporate them into living cells again. Thus if the cells are removed, the mineral content of the treated water is reduced and its ability to cause eutrophication of receiving waters is greatly diminished. On the other hand, if algal cells are not harvested, the pond itself rapidly becomes a highly eutrophied body of water although its effluent is as highly stabilized as is possible with any normal engineered system.

When the low cost of the stabilization pond is considered along with its ability to upgrade the quality of water, it is evident that such a system is a device for water quality management worthy of engineering attention, particularly where land areas and geographical and climatological relationships are favorable.

GENERAL CONCEPT
OF STABILIZATION PONDS

The term "stabilization pond" is generally applied to artificially created bodies of water intended to retain waste flows containing degradable organic compounds until biological processes render them stable and hence either unobjectionable from an oxygen-demand viewpoint for discharge into natural waters, or are removed by percolation and evaporation. The theoretical minimum time of such retention is that sufficient to permit biodegradation of organic matter and die-away of pathogenic bacteria and parasites. The theoretical maximum time, from a quality control viewpoint, is the minimum time plus that necessary to tie up the stable products of biodegradation in algal cells. In practice, however, the design detention period, or pond capacity, may be governed by requirements of controlled discharge imposed by quality management considerations of the receiving water.

TYPES OF
STABILIZATION PONDS

Stabilization ponds may be classified according to types of influent, outflow condition, methods of oxygenation, biological processes, etc. Typical general classes are as follows:

1. *Types of influent*
   (a) *Raw sewage ponds, or lagoons* Influent sewage is discharged directly from municipal sewer without treatment.
   (b) *Screened sewage ponds* Influent sewage is screened or comminuted.
   (c) *Primary sewage ponds, or oxidation ponds* Influent sewage is effluent from primary sedimentation system.

    (*d*)   *Secondary sewage ponds*   Influent sewage is effluent from secondary sewage treatment system.

2. *Outflow conditions*
    (*a*)   *Percolation beds*   Outflow by evaporation and percolation in soil exceeds influent rate.
    (*b*)   *Nonoverflow ponds*   Outflow by evaporation and percolation equals influent plus precipitation.
    (*c*)   *Intermittent ponds*   No outflow during dry season but effluent discharge during wet periods.
    (*d*)   *Overflowing ponds*   Effluent discharged continuously.

3. *Methods of oxygenation*
    (*a*)   *Mechanical aeration*   Water aerated essentially continuously by brushes, low-head propeller pumps, floating or submerged aerators, etc. (Principles of "total oxidation" activated sludge apply; sludge must be withdrawn.)
    (*b*)   *Photosynthetic oxygenation*   During growth and photosynthesis, algae produce dissolved oxygen in amounts equal to or greater than the oxygen demand (BOD) of the applied waste.

4. *Biological processes*
    (*a*)   *Aerobic ponds*   Loaded so that aerobic conditions prevail and biological processes are mainly bio-oxidation and photosynthesis.
    (*b*)   *Anaerobic ponds*   Loaded to such extent that anaerobic conditions prevail through water mass. Biological processes primarily organic acid formation and methane fermentation.
    (*c*)   *Facultative ponds*   Divided by loading and thermal stratification into aerobic surface and anaerobic bottom strata.

From the foregoing summary, it is evident that any single pond may fall into several classifications. For example, a pond may be an aerobic primary or secondary pond, overflowing or intermittent, etc. On the other hand, because it is a quality control device, it may not be a raw sewage nonoverflow pond.

FUNDAMENTAL PRINCIPLES
OF STABILIZATION PONDS

Although stabilization ponds are physically nothing more than simply constructed shallow (5 to 6 ft) earthwork basins, their effectiveness depends upon a complex interaction of physical, chemical, and biological processes. In natural ponds, however, these processes proceed without the watchful eye of highly competent chemists and microbiologists. Moreover, a minimum amount of engineering, construction, and mechanical equipment is required. These factors, together with the tendency for industrialized nations to utilize mechanized treatment processes discussed in Chapter 14, long diverted scientific attention away from the stabilization pond as a practical water quality control system. Engineering attention began largely in the context of providing a cheap substitute for sewage treatment for communities too poor to afford modern works, but as the merit of ponds became evident, attention was turned to the underlying principles and parameters of design.

One of the first fundamental principles underlying the most used types of stabilization ponds is that their action depends upon the simultaneous and continuous functioning of both the right-hand and the left-hand sectors of the aerobic cycle of organic growth and decay (Figure 2-2). This principle contrasts with the conventional system which carries out only the degradation process and leaves the growth potential to be exerted in the receiving water, as described in Chapter 11 (eutrophication, etc.). Figure 17-1 illus-

Figure 17-1 *The cycle of photosynthetic oxygenation.*

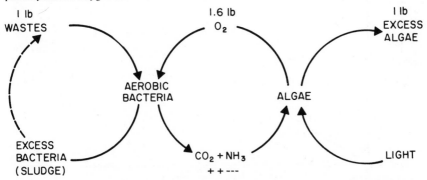

trates the synergistic activity of bacteria and algae in photosynthetic oxygenation. Significantly, although the system is internally self-sufficient, the input is biodegradable dead organic wastes and the output is living organic matter at a higher energy level. Hence the potential oxygen demand (BOD) of the effluent may be greater than that of the influent, provided of course that the organic waste is sewage or other partially degraded organic matter. The living algal cells, however, are not quickly available for biodegradation because of their tenacity of life. Nevertheless, in terms of water quality, the stabilization pond effluent may substitute an aesthetic factor for the quality factors associated with biodegradation unless the algal cells are harvested.

To express these two factors in more precise terms, it may be said that the waste matter entering the system has some definite BOD which, when satisfied through biodegradation, produces stable products having an algal growth potential (A.G.P.). *A.G.P. may be defined as the weight of algae which will grow at the expense of algae nutrients in a water when no factor other than nutrient is limiting to growth.*

The relationship between BOD and A.G.P. for raw domestic sewage and for completely oxidized effluent is illustrated in Figure 17-2. A.G.P., whether it be generated in a stabilization pond or in a conventional system, must be considered a factor in water quality. The significant fact is that it may have to be controlled either by harvesting algae from the pond effluent or by nu-

Figure 17-2   *Relationship between A.G.P. and BOD.*

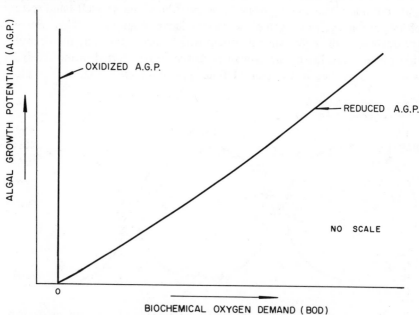

ALGAL GROWTH POTENTIAL (A.G.P.)

OXIDIZED A.G.P.

REDUCED A.G.P.

NO SCALE

O

BIOCHEMICAL OXYGEN DEMAND (BOD)

trient removal from the conventional secondary plant effluent, neither of which is economically acceptable at the present time. A further factor to be considered in relation to the water quality effects of stabilization ponds or conventional works is the possibility of growth stimulation by effluents from which nutrients have been removed. That is to say, effluents which in themselves do not contain nitrogen or phosphorus may contain other factors (trace elements, vitamins, etc.) capable of triggering eutrophication of receiving waters which contain nutrients but are limited by certain growth factors in their ability to support life. Hence quality management may involve productivity tests not commonly utilized in the past.

The simultaneous processes of growth and decay in stabilization ponds are affected by numerous environmental factors. Insolation and radiation, temperature and thermal gradients, pond geometry, wind, gas exchange, and seeding are among the most important which have been evaluated experimentally [3]. The biochemical activity involves typical carbon, nitrogen, sulfur, and phosphate transformations, as well as more subtle reactions, all of which vary with the type of pond. Principal reactions are summarized in Table 17-1, and their interactions shown schematically in Figure 17-3.

Figure 17-3 *Interrelationship of conditions and reactions in ponds.*

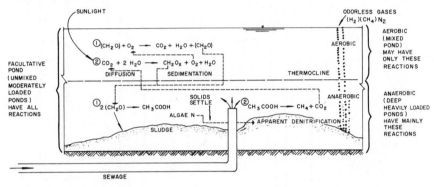

ENGINEERING DESIGN CONSIDERATIONS [4, 5]

A number of considerations enter into the design of a stabilization pond. In general they concern the physical integrity of the structure or the determination of pond size. For example, a nonoverflow pond must be of adequate size to dispose of the influent by a combination of evaporation and percolation at all seasons of the year. A mass diagram, based on anticipated influent rates, seasonal rainfall, local evaporation rates throughout the year, and equilibrium percolation rates, can be utilized in the normal engineering fashion to establish the surface area and embankment freeboard requirements. Most of these components can be readily estimated from normally available data. Percolation rates, on the other hand, are not so easily determined. As noted in Chapter 16, the equilibrium rates for most soils may be quite small. In some small stabilization ponds in California, percolation rates of 0.3 in./day have been observed. Table 17-2 gives data for combined evaporation and percolation for some small ponds in central California.

Based upon currently available data, a conservative rule of thumb for estimating purposes in average soil in semiarid climates is 10,000 to 15,000 gal/acre/day outflow via percolation and evaporation combined.

Design parameters which must be established for stabilization ponds include detention period, hydraulic loading, organic loading, depth, recirculation, mixing, pond size and shape, and inlet and outlet systems.

TABLE 17-1 *Summary of principal biochemical transformations in ponds*

Type of pond	Reaction	Organisms involved	Typical reaction
Carbon transformations			
Aerobic	Biological oxidation	Aerobic bacteria, fungi, and protozoa	$(CH_2O)_x + xO_2 \rightarrow xCO_2 + xH_2O$
	Photosynthetic oxygenation	Algae: *Chlorella, Scenedesmus, Euglena,* and various spp.	$CO_2 + 2H_2O + h\nu \rightarrow (CH_2O) + O_2 + H_2O$
Anaerobic	Organic acid formation	Facultative heterotrophs	$2(CH_2O)_x \rightarrow xCH_3COOH$
	Methane fermentation	Methane bacteria	$CH_3COOH \rightarrow CO_2 + CH_4$
Nitrogen transformations			
Aerated			Organic N $\rightarrow$ Ammonia N $\rightarrow$ Nitrate N $\rightarrow$ Denitrification
Aerobic			Organic N $\rightarrow$ Ammonia N $\rightarrow$ Algae N (removed)
Facultative			Organic N $\rightarrow$ Ammonia N $\rightarrow$ Algae N $\rightarrow$ Inorganic N (?)
Anaerobic			Organic N $\rightarrow$ Ammonia N

Sulfur transformations

Aerobic		Organic sulfur → sulfate
Anaerobic	Photosynthetic bacteria: *Thiopedia Chromation*	Organic $S \rightarrow HS^- + H^+ \underset{\text{Basic}}{\overset{\text{Acid}}{\rightleftharpoons}} H_2S \uparrow$ $2H_2S + CO_2 + h\nu \rightarrow (CH_2O) + S_2 \downarrow + H_2O$

Phosphate transformations

Aerobic		Organic $P \rightarrow H_3PO_4 \rightarrow$ Calcium phosphate $\downarrow$
Anaerobic	Phosphate reduction	Organic $P \rightarrow (?)$

TABLE 17-2 *Percolation plus evaporation rates as a function of soil types*

Pond system observed	Pond size, acres	Pond depth, in.	Soil type	Approximate terminal rate*	
				In./day	Gal/acre/day
Woodland South	2	60	Heavy silty clay	0.40	11,000
Esparto					
July, August	2	72	Light silty clay	0.66†	18,000
February, March	2	72	Light silty clay	0.25‡	6,800
Woodland North	2	60	Alkaline silt	0.73	20,000
Woodland Pilot	¼	60	Alkaline silt	0.75	20,200
Gonzales	3	72	Fine sand	1.3	35,000
Rio Lindo	½	72	Gravel with silt	1.5	41,000

*Includes evaporation, which may average 0.1 in./day, and may equal 0.3 in./day during summer. Observed percolation rates apparently decrease in winter and increase slightly during summer.

†July, August evaporation = 0.33 in./day, therefore percolation = 0.33 in./day.

‡In California evaporation and precipitation are usually about equal in March, therefore net loss is primarily due to percolation alone.

Detention period

The detention period necessary to accomplish the function of a stabilization pond depends, of course, upon the type of pond and hence the nature of the reactions upon which its action depends; the type of waste applied; and various environmental conditions. A summary of such characteristics and environmental factors for various types of ponds receiving domestic return flows is presented in Table 17-3.

In a pond of depth d, length l, and width w, the detention period D is described by the simple equation:

$$D = \frac{V}{Q} = \frac{wld}{Q}$$

where V is volume and Q is the influent rate.

Selection of the appropriate detention period for various types of ponds is subject to the following considerations:

1. *Aerobic ponds*   Photosynthetic oxygen production is equated to the oxygen demand of the waste. A determinant time is required for accumulation by algae of sufficient sunlight energy to liberate the required oxygen, according to the following equation [4]:

$$\frac{d}{D} = \frac{0.66FS}{L_a}$$

where $F =$ oxygenation factor (Figure 17-4)
$S =$ solar radiation (Table 17-4)
$d, D,$ as previously noted
$L_a =$ first stage BOD
For best light utilization the required detention period rarely exceeds 5 days.

2. *Facultative ponds*   Detention period is selected to provide either suitable time for coliform die-away or sufficient time to permit combined evaporation and percolation to equal the influent volume, depending upon which of the two is required by pollution control considerations. It is seldom that the computed detention period is so short that the removal of suspended solids or BOD is hampered by lack of time. However, even though detention periods may be as long as 120 days, effluent BOD rarely falls below 10 ppm.

3. *Anaerobic ponds*   Detention period is selected to permit complete anaerobic breakdown without diminishing the organic load rate. To assure anaerobic conditions, organic load should exceed 200 lb/acre/day. Detention periods are best made long by providing as much depth as is feasible.

TABLE 17-3 *Summary of characteristics and environmental requirements of the major biological reactions in waste disposal ponds*

Biological reaction	Characteristics				
	Organisms	Usual substrate	Major products	Time required, days*	Odors produced
Aerobic oxidation	Aerobic bacteria	Carbo-hydrates, proteins	$CO_2 + NH_3$	5-10	None
Photosynthetic oxygenation	Algae	CO_2, HN_3	Oxygen, algae	10-20	None
Acid formation	Facultative heterotrophic bacteria	Carbo-hydrates, proteins, fats	Organic acids	10-20	H_2S Organic acids
Methane fermentation	Methane producers	Organic acids	CH_4, CO_2, H_2	40-50	H_2S

Biological reaction	Environmental factors				
	Temperature, °C (P/O)†	Oxygen	pH	Light	Toxic compounds
Aerobic oxidation	$\frac{0\text{-}40}{15\text{-}30}$	Required	7.0- 9.0	Not re-quired	Cr^{+++}, NH_3+
Photosynthetic oxygenation	$\frac{4\text{-}40}{15\text{-}25}$	Required under certain conditions	6.5-10.5	Required	Ca^{++}, Cl_2, Cr^{+++}
Acid formation	$\frac{0\text{-}50}{10\text{-}40}$	Required under certain conditions	4.5- 8.5	Not re-quired	Cr^{+++}, Cl_2
Methane fermentation	$\frac{6\text{-}50}{14\text{-}30}$	Must be excluded	6.8- 7.2	Not re-quired	O_2, deter-gents

*Time required after first initiation of pond to develop a stable population.
†P/O = (permissible range)/(optimum range).

Hydraulic loading

Hydraulic loading (u) is normally expressed in terms of inches per day, therefore it is equal to Q/A; or, in terms of detention period,

$$u = \frac{d}{D}$$

Figure 17-4 *Relationship between oxygenation factor and BOD removal in waste ponds.*

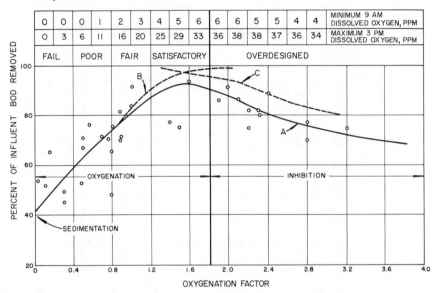

Typical values of hydraulic load for various types of ponds are:

1. *Aerobic ponds*　　Usually limited to 2 to 10 in./day.
2. *Facultative ponds*　　Usually limited by overflow requirements. Overflow may be equated to evaporation and percolation for nonoverflowing systems. Maximum recommended hydraulic load is about 1.5 in./day for domestic waste.
3. *Anaerobic ponds*　　Value should be selected to give 40 or more days of detention period to permit complete digestion.

Organic loading

Organic loading (L_o) is expressed in pounds of BOD per unit area of pond surface per day (lb/acre/day). It is proportional to the hydraulic load and the organic concentration or ultimate BOD (L_a). Therefore

$$L_o = 0.226 \frac{d}{D} L_a$$

TABLE 17-4 *Probable values of visible solar energy as a function of latitude and month*

Latitude, deg N or S	Month											
	Jan	Feb	Mar	Apr	May	Jun	Jul	Aug	Sep	Oct	Nov	Dec
0 max	225*	266	271	266	249	236	238	252	269	265	256	253
min	210	219	206	188	182	103	137	167	207	203	202	195
10 max	223	244	264	271	270	262	265	266	266	248	228	225
min	179	184	193	183	192	129	158	176	196	181	176	162
20 max	183	213	246	271	284	284	282	272	252	224	190	182
min	134	140	168	170	194	148	172	177	176	150	138	120
30 max	136	176	218	261	290	296	289	271	231	192	148	126
min	76	96	134	151	184	163	178	166	147	113	90	70
40 max	80	130	181	181	286	298	288	258	203	152	95	66
min	30	53	95	125	162	173	172	147	112	72	42	24
50 max	28	70	141	210	271	297	280	236	166	100	40	26
min	10	19	58	97	144	176	155	125	73	40	15	7
60 max	7	32	107	176	249	294	268	205	126	43	10	5
min	2	4	33	79	132	174	144	100	38	26	3	1

*Values of solar radiation S in cal/cm²/day.

To determine average value of S: $S_{avg} = S_{min} + p(S_{max} - S_{min})$ in which p is total hours sunshine divided by total possible hours sunshine.

To determine yield of algal cell material, in lb/acre/day: $Y_c = 0.15\ FS$.

To determine yield of oxygen, in lb/acre/day: $0_2 = 0.25\ FS$.

in which the constant 0.226 is the conversion factor from mg/liter BOD to lb/acre/day.

Typical organic loadings for various pond types are:

1. *Aerobic ponds*   Maximum load is about 200 lb of ultimate BOD per acre per day, but load is dependent upon available sunlight energy [5]. A first approximation is

$$L_o = S$$

in which L_o is the ultimate BOD to be satisfied in lb/acre/day, and S is the quantity of sunlight energy in cal/cm^2/day. (See Table 17-4.)

2. *Facultative ponds*   Maximum load is based upon mass diagram for gas evolution, since evolution of methane is the major method of carbon escape. Mass diagram of gas, evolved from a pond in Woodland, northern California, indicated an average rate of organic loading of 50 lb of ultimate BOD per acre per day. If 50 lb of BOD per acre per day is exceeded continuously, it is believed that the effects of eutrophication will become evident within 3 to 5 years.

3. *Anaerobic ponds*   Permissible loading is a function of rates of methane evolution and organic overflow. Maximum BOD removal in anaerobic ponds is about 70 percent. Therefore anaerobic ponds must normally discharge to secondary ponds.

Depth

Considerations in the selection of pond depth and the normal values used are:

1. *Aerobic ponds*   Algae concentration is an inverse function of depth; light conversion efficiency increases with depth. Yield of algae or oxygen is a product of efficiency and concentration. Maximum yield and concentration occur with pond depth held at 6 to 12 in. However, if oxygen production is the sole objective, depths of 36 to 48 in. are feasible provided mixing is applied.

2. *Facultative ponds*   Selected depth should be such that the summer thermocline occurs several feet above the pond bottom to provide a sufficiently deep stratum in which sludge accumulation and digestion can take place below the thermocline where oxygen is unlikely to intrude. This criterion is usually best met by maintaining depths of 60 to 72 in. Ponds as deep as 96 in. are observed to operate satisfactorily. A cone-shaped central well surrounding a central inlet forms an excellent digester. Cone depths of 3 or 4 ft. are used.

3. *Anaerobic ponds*   Depth is selected to provide minimum surface-area-to-volume ratio for heat retention and to provide a detention period for alkaline digestion. Criteria are usually best met by maintaining depths of 96 to 168 in. Maximum depth is limited by construction cost and

the possibility that excessively low temperatures will prevail at the pond bottom, precluding methane fermentation.

Recirculation

Recirculation of the contents of stabilization ponds is necessary in some cases, and not in others.

1. *Aerobic ponds*   In high-rate ponds recirculation is desirable because it provides influent dilution, oxygenation, and seeding. A recirculation rate of at least Q is desirable.

2. *Facultative ponds*   Recirculation is mainly justified if edge or corner inlet is used, if ponds are long and narrow, or if loading in primary ponds of a series exceeds the maximum at which aerobic surface conditions can be maintained. Should recirculant be mixed with incoming sewage the anticipated temperature of recirculant should be considered in calculations so as to avoid cooling the influent sewage to temperatures below that needed for digestion.

3. *Anaerobic ponds*   Recirculation of actively digesting sludge from the anaerobic zone into income waste is feasible provided there is no entrainment of oxygen into the digesting bottom layer. Submerged application is recommended for the prevention of the escape of odors.

Mixing

1. *Aerobic ponds*
    (a)   Mixing is essential for the commingling of the settleable bacterial sludge and the algae which remain in the supernatant. Mixing may be accomplished by flowthrough, using airlift pumps, propeller pumps, brush aerators, or various other methods. Around-the-end baffles with propeller pumps is believed to be most economical for photosynthetic oxygenation.
    (b)   Mixing systems should create a velocity of at least ½ ft/sec throughout the matrix of the culture.
    (c)   Mixing is best applied
        (1)   at night during the period when dissolved oxygen is at a level below saturation (usually between midnight and 5 A.M.).
        (2)   when the pH of the culture exceeds 9.5, so that CO_2 may be replenished in the supernatant.

2. *Facultative ponds*
    (a)   Mixing is not required for facultative ponds to which the BOD loading is 30 to 60 lb/acre/day.
    (b)   In aerated facultative ponds mixing should be applied only to the aerobic zone, and then mainly at night.

3. The effect of mixing on anaerobic ponds has not been studied, but it is obvious that aeration with oxygen must be avoided since oxygen is toxic

to the methane bacteria. Some mixing is naturally provided by rising gas bubbles when vigorous fermentation is in progress. Vigorous gasification prevents formation of thermocline.

Pond size and shape

1. *Aerobic ponds (photosynthetic oxygenation)*   Maximum size of individual ponds should be about 10 acres. (Size is governed by depth, head loss during mixing, and channel width.) An extensive mixing facility is required for channels which are in excess of 50 ft wide. Power for mixing is about 3 kwhr/acre/day. A length-to-width ratio of 4:1 is optimum for overall pond. Channel lengths of 10,000 ft are not uncommon.

2. *Facultative ponds*
   (a) *Overflowing ponds*   No size limitation other than cut and fill balance consideration in earthwork. However, small ponds give superior performance since lateral percolation is maximized, and problems of wind mixing and erosion are minimized.
   (b) *Nonoverflowing ponds*   Percolation of liquid into levees may dispose of significant quantities of effluent into the ground. Individual ponds as small as ¼ acre are feasible. Two-acre square ponds arranged in tetrads is an excellent configuration. In square ponds, corner radii should not be less than 40 ft.

3. *Anaerobic ponds*
   (a) *Overflowing ponds*   From ½ to 2 acres is common. Length-to-width ratios of from 2:1 to 4:1 will provide inlet and outlet separation.
   (b) *Nonoverflowing ponds*   Size should be selected to minimize surface-area-to-volume ratio, thereby conserving heat. May be square-shaped with rounded corners.

Inlet and outlet arrangements

1. *Aerobic ponds*   Arrange to avoid short-circuiting during periods when mixing is not in progress. Inlet is usually on discharge side of mixing system; outlet on intake side of mixing system. Check valves are used on pump outlets to avoid backflow. Outlet is arranged to decant overflow.

2. *Facultative ponds*
   (a) *Inlets*   Central inlets are superior to all others. Deposited sludge will then be entirely submerged and sludge will be protected from aeration. Inlet pipe should be turned up and extended about 18 in. to avoid possibility of clogging. See Figure 17-5a.
   (b) *Transfer structures*   Simple pipes through the levee are satisfactory provided erosion control at the discharge is used. A supe-

rior transfer structure uses the decanting and submerged discharge principle. A typical system is shown in Figure 17-5*a*.

    (*c*)   *Outlet*   Should be located downwind so that outlet skims pond. Initial and final ponds in series may have submerged discharge to avoid transfer of grease or discharge of scum to receiving water. See Figure 17-5*b*.

3.   *Anaerobic ponds*   Same as facultative ponds except that inlet pipe should extend upward about 5 ft to avoid clogging. (See Figure 17-5*c*.)

Figure 17-5

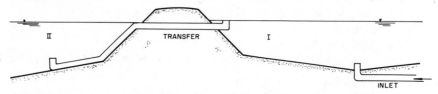

A. INLET AND SURFACE TRANSFER STRUCTURE

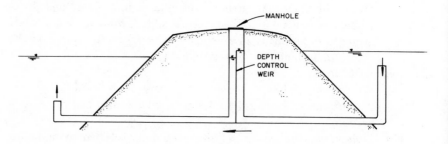

B. SUBMERGED EFFLUENT OR TRANSFER STRUCTURE

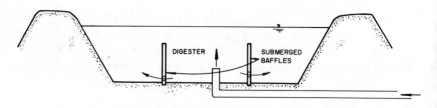

C. TYPICAL USE OF SUBMERGED BAFFLES TO PROVIDE DIGESTER

System arrangements

The arrangements of ponds into systems of various types is diagramed in Figure 17-6 in which flow patterns are shown for single ponds and for

groups of two, three, and four ponds. Parallel ponds give maximum load distribution, whereas series ponds have the advantage of producing superior effluents [6].

Miscellaneous design considerations

1. Pond linings

    (*a*) *Aerobic ponds* Rapidly mixed photosynthetic oxygenation ponds should be lined if most effective oxygenation is to be at-

Figure 17-6 *Flow patterns for waste stabilization systems.*

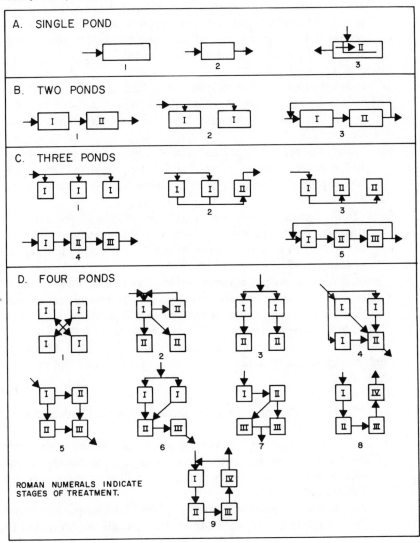

tained. Lining is not essential if the velocity is low, although mixing may cause undue turbidity, resulting in increased light adsorption and severe light limitations for the algae. Linings may be of asphalt, gunite, or plastic. Galvanized sheet iron forms a suitable lining for small aerobic ponds, since carbon dioxide removal by algae causes the pond water to have a positive Langelier index with deposition of insoluble substances on the metallic surfaces [7]. Deep aerobic ponds need not be lined if scouring velocity is not exceeded, that is, if velocity is below ½ ft/sec.

(b) *Facultative and anaerobic ponds*   Linings are not required for the fostering of any fundamental reactions. Linings may be applied to prevent excessive rates of percolation or bank erosion. Cost of linings is high because of the large pond areas involved. Liners may be of plastic, clay, or compacted earth.

2. *Levee slopes*

(a) *Aerobic ponds*   Levees need not be high, and since they should be lined, they may be constructed with slopes sufficiently flat to permit operation of paving machines. Recommended minimum slopes: 4:1.

(b) *Facultative ponds*   Recommended interior levee slopes, 4:1 to 6:1; exterior slopes, 3:1. Waterline may be paved to prevent beaching. Riprap should not be used as a substitute for shallow slopes.

3. *Levees*

(a) *Width*   Since levees will provide percolation in proportion to their size, design size should be generous. Recommended minimum top width: 14 ft. Levee crown should be rounded to prevent impoundment of rainwater.

(b) Redwood or precast concrete baffles may replace inner levees in high-rate ponds or in overflowing ponds when water surface area must be kept at a maximum and land is at a premium. Submerged baffles may be used to create an anaerobic zone. (Cf. Figure 17-5c.)

4. *Freeboard*   Minimum recommended freeboard, 2.0 feet. Maximum freeboard governed by anticipated wave size (usually one-half the side-water depth).

5. *Appurtenances*

(a) *Meters*   Provision should be made for inlet and outlet metering in every ponding system.

(b) *Distribution systems*   These systems usually are open boxes containing valves to control flow through pipes leading to the various ponds. For primary ponds rotary distributors give equalized load distribution.

(c) *Facultative and anaerobic ponds*   Single or multiple submerged vertical inlet pipes should be used. Vertical inlet height should be about 18 in. above pond bottom. Primary pond inlet should be at center of pond.

(d) *Individual pond inlets*   There are no special inlet requirements for aerobic ponds except to avoid short-circuiting of influent sewage directly to effluent overflow.

(e) *Drains*   Each pond should be provided with a drain sump, with drainage pipes leading to the influent well. Ponds may thus be emptied into one another when one is to be removed from operation. Thus the key phrase "keep ponds full or empty" can be accomplished. Portable pumps and portable irrigation piping may be provided in lieu of permanent drains and pump installations.

(f) *Skimmers*   Each primary pond should be provided with a surface skimming device located within a few feet of the levee-waterline intersection at the downwind edge of the pond. Skimmers should be adjustable and operable manually, and should drain to the influent or recirculation sump.

(g) *Outlet and interpond transfer systems*   These systems may be combined with pond drains or may be designed to operate as simple culverts through levees. In freezing climates the submerged transfer or overflow lines should be connected through a manhole containing a depth control weir.

Management of ponds

In a preceding section attention was called to the fact that the ability of a natural pond to function without scientific personnel led to an ignoring of the pond as a treatment device. This does not mean that the engineered pond functions without attention and maintenance. Such an assumption essentially degrades all ponds to the status of a severely polluted pothole with attendant objectionable aesthetic conditions. Such neglect has repeatedly occurred and has done much to obscure the merits of the engineered pond in some quarters. The functioning of an aerated pond, an aerobic pond, a facultative pond, or an anaerobic pond depends upon management techniques which maintain optimum conditions for the processes which the pond is intended to exploit. Proper operation and maintenance, therefore, are just as essential to the stabilization pond as to other engineered systems for quality management. It follows, therefore, that conscientious and intelligent operation and maintenance of ponds cannot be justifiably ignored under any circumstances.

PERFORMANCE OF STABILIZATION PONDS

The effects of stabilization ponds on the quality of domestic return flows have been the subject of many experiments. BOD reductions which may be expected from various types of ponds are as follows:

1. *Aerated ponds*   Up to 95 percent conversion to CO_2 and humus
2. *Aerobic ponds*   Up to 95 percent conversion to algae and humus
3. *Facultative ponds*   90 to 95 percent conversion to algae, methane, and humus
4. *Anaerobic ponds*   Up to 70 percent conversion to methane and humus

Conversion of BOD to algal cell material in aerobic ponds produces about 1 ton of algae per million gallons of domestic sewage ponded. The short detention period involved results in but small reduction of the coliform count. Facultative ponds, however, are more effective in reducing bacteria. A single pond at Santee, California, has been observed to remove 50 percent of influent coliforms. A series of facultative ponds providing an overall detention period of 60 days produce water of a quality suitable for golf course irrigation and landscaping. Coliform bacteria counts in the third of a series of ponds receiving domestic sewage are generally less than 10 per 100 ml. Fish, frogs, and invertebrate life likewise attest to good quality of water in tertiary ponds.

Observations [8] of pesticides in aerobic ponds show that many are adsorbed on algal surfaces. In preliminary experiments with algal harvesting, 80 percent of pesticides were removed.

Degradation of detergents in stabilization pond studies [9] are summarized in Table 17-5.

TABLE 17-5   *Removal of detergents in stabilization ponds*

Type of pond	Percent removal (R) and degradation (D)					
	ABS		LAS		Alcohol sulfate	
	R	D	R	D	R	D
Aerobic	15	2.9	56.2	35.3	95.3	94.2
Facultative	30	15.9	93.1	80.5	98.7	94.0
Anaerobic	Not observed. Other available systems ineffective.					

STATUS OF STABILIZATION PONDS

Stabilization ponds are now employed to treat municipal return flows, and to retain or treat many types of industrial waste waters, throughout the United States and most civilized countries of the world. For example, in 1957 there

231

were some 681 municipal ponds in the United States [10]. By 1960 the estimated number was above 1,000 [11]. Although data for the entire United States were not compiled in 1966, there were more than 1,200 municipal and industrial ponds in California alone, with new ones reported every few weeks. At Melbourne, Australia [12], ponds covering more than 500 acres are a major factor in treating the waste water from domestic use. Numerous ponds are in use in India and the Middle East. In Israel more than 75 stabilization ponds were in operation in 1965 [13], and plans are well advanced [14] to utilize ponds in the treatment and reclamation of water from the city of Tel Aviv, and the entire Dan Region, in the amount of some 90 million gpd.

The range of applicability of stabilization ponds has not been fully explored. Domestic sewage, food-processing wastes, animal manures, oil, and certain mineral wastes are among the water-carried wastes which have been treated successfully. However, the feasibility of utilizing ponds for the biodegradation of solid organic wastes in a pond seems worthy of exploration. In such ponds there would be no overflow and the organic matter would be reclaimed by conversion to algal cells which could be harvested for animal feed supplements. In any event, for maximum quality objectives of the system as a water-treatment device, provision should be made for removal of algal cells, although such removal is not implicit in the concept of the stabilization pond and has not yet been established on an economic basis.

Nevertheless, the design of stabilization ponds is rapidly being placed on a rational basis, and their usefulness as engineered systems for water quality management is certain to grow in importance as the objectives of such control become increasingly restrictive.

REFERENCES

1 C. G. Gillespie: "Disposal of Liquid Industrial Wastes from the Viewpoint of the Bureau of Sanitary Engineering of the California State Department of Public Health—A Look to the Future," *Sewage Works J.*, 16(6):956, 1944.

2 G. P. Fitzgerald and G. A. Rohlich: "An Evaluation of Stabilization Pond Literature," *Sewage Ind. Wastes*, 30(10), 1958.

3 W. J. Oswald: "Fundamental Factors in Stabilization Pond Design," *Advances in Waste Treatment*, Pergamon Press, New York, 1963.

4 W. J. Oswald: "Advances in Stabilization Pond Design," *Proceedings of the 3d Annual Sanitary Engineering Conference*, Vanderbilt University, Nashville, Tennessee, 1964.

5 W. J. Oswald: "Light Conversion Efficiency of Algae Grown in Sewage," *Trans., ASCE*, 128(III), 1963.

6 G. V. R. Marais and V. A. Shaw: "A Rational Theory for the Design of Sewage Stabilization Ponds in Central and South Africa," *The Civil Engineer in South Africa*, 3(11), 1961.

7 C. G. Golueke et al.: "Nutritional and Disease Transmitting Potential of Sewage Grown Algae," *Second Progress Report* (WP 00026), Sanitary Engineering Research Laboratory, University of California, Berkeley, 1964.

8 Donald S. Crosby: Department of Toxicology, University of California, Davis, private communication, 1964.

9 S. A. Klein: "Fate of Detergents in Septic Tank and Oxidation Pond Systems," supplement to University of California SERL Rept. 64-1, May, 1964.

10 *Manual of Engineering Practice*, American Society of Civil Engineering, 36, 1959.

11 Anon.: *Proceedings of the 1961 Kansas City Conference on Waste Ponds*, U.S. Public Health Service, Kansas City, 1961.

12 C. D. Parker: "Microbiological Aspects of Lagoon Treatment," *J. Water Pollution Control Federation*, 34(2), February, 1962.

13 A. Meron, M. Rebhun, and B. Sless: "Quality Changes as a Function of Detention Time in Wastewater Stabilization Ponds," *J. Water Pollution Control Federation*, 37(12), December, 1965.

14 A. Amramy, B. Caspi, and A. Melamed: "Dan Region Sewage Reclamation Project, General Project Outline," P.N. 241, TAHAL (Tel-Aviv), Water Planning For Israel Ltd., September, 1962.

CHAPTER 18

Quality management
by dilution in freshwater

INTRODUCTION

The historic concept of sewerage as a problem in transportation of wastes from the household to a flowing stream, noted in a previous chapter, led quite logically to a consideration of the stream as a treatment system in itself. Initially, perhaps, disguising of the wastes by simple dilution may have been sufficient, but as the ratio of waste water to freshwater increased it was no longer possible to ignore the presence of wastes if the stream was to serve any beneficial use beyond that of wastes transport.

Several factors make necessary an evaluation of the ability of freshwater systems to serve as waste treatment works. Among the important areas are the following:

1. Domestic use of water, being largely nonconsumptive, involves return flows which must generally go back to the freshwater resource.
2. The technology of waste-water treatment is not adequate to erase all traces of domestic or industrial use.
3. Economic concepts tend to minimize investment in waste-water treatment.
4. The economic need of cities for industry and of industries for water supply has in the past led to permissiveness in the matter of waste discharges, at least within the limits of public tolerance of the results.
5. Sanitary standards increase progressively with urbanization in an expanding economy.

Thus scientific and engineering effort has been directed to an understanding of what occurs when wastes are discharged into the freshwater resource, as well as how to predict the quality factors which may go into a stream with return flows under the limitations imposed by any given public policy of quality control.

233

Ponds and lakes as well as streams may, of course, serve as treatment works. Nevertheless, the flowing stream has been given most attention. Public disinterest in its own wastes, however, has made for slow progress in such effort.

THE STREAM AS A
WASTE TREATMENT SYSTEM

It is only necessary to recall that in nature organic matter will be degraded in accordance with the aerobic cycle if oxygen is adequate, or by a sequence of anaerobic and aerobic cycles if it must, to anticipate what will occur when domestic wastes are discharged to a stream. And since a waste-water treatment system is simply a device to utilize these cycles under optimum conditions, a flowing stream must be considered as one type of treatment works.

Fundamentally a freshwater stream differs from a more conventional engineered treatment works principally in that:

1. It may be a hundred miles long instead of a few hundred feet.
2. Its biological processes occur sequentially rather than simultaneously in a single location.
3. It is subject to less precise design parameters.
4. Optimum environmental conditions for biodegradation are more difficult to maintain.
5. It interferes with other normal human activity.

Significantly, the concept of a stream as a waste treatment system has been recognized in legislation. Many water quality control laws (e.g., California) specifically recognize waste disposal as one of the beneficial uses which quality control agencies may choose to protect in any individual situation. Generally, however, this use is recognized to the extent that it is compatible with other beneficial uses and quality control objectives, but in few cases is it feasible to rule it out entirely. Consequently, there is need for both engineering parameters by which to judge the waste-assimilating potential of a stream, and criteria by which to relate water quality to management purposes.

The following are typical objectives and rough criteria which have long been in common use:

1. *Purpose*   Protection of water supplies.
    Criteria   Not more than 5,000 coliform organisms per 100 ml, with occasional maxima up to 20,000 per 100 ml; no toxic chemicals or chemicals which give water objectionable taste, odor, etc.
2. *Purpose*   The preservation of useful aquatic life.
    Criteria   (Ellis [1]); dissolved oxygen content of 5 mg/liter minimum, for good mixed fish faunae; occasional short period decline to

4 mg/liter permissible; absence of bottom deposits to disrupt food chain, or of toxic materials, or excess turbidity which excludes sunlight.

3. *Purpose*   The prevention of nuisance.
   Criteria   Absence of unsightly sludge deposits, floating material, and odor. Generally 3 to 6 cfs streamflow per 1,000 population will disguise sewage as far as senses of sight and smell are concerned.

4. *Purpose*   The protection of bathing beaches.
   Criteria   No observable floating solids; not more than 10 coliform organisms per 1.0 ml. (Note: No satisfactory criteria for bathing beaches exist.)

5. *Purpose*   Protection of structures.
   Criteria   Absence of acids and other products of decomposition which attack metal ships or ship paint; absence of H_2S or other gases deleterious to house paint and metals.

Of these several purposes the one most considered from an engineering approach is that of maintaining an adequate oxygen resource in the receiving water. Few instances are to be found in the United States today where simple prevention of nuisance is adequate to the quality goals of public policy; and for almost all other purposes the limiting conditions prescribed cannot be exceeded in a water having adequate oxygen. Thus *the dissolved oxygen criterion requires that an aerobic cycle of biodegradation be maintained in the stream and that the concentration of BOD shall be limited.*

Before considering how such BOD limits may be estimated, let us first consider what can occur when an untreated domestic return flow is discharged into a normal stream carrying dissolved oxygen and supporting an aquatic society which ranges from bacteria to plankton to fly larvae to minnows to fishes; which has rapids and pools and otherwise is subject to reaeration; and which does not have a flow of such exceedingly great magnitude that dilution readily takes care of the pollution problem. What takes place has arbitrarily been divided into four zones, summarized as follows, and shown graphically in Figure 18-1. (See also Figure 9-1.)

1. *The zone of degradation*   In this zone decomposing sewage begins to depress the oxygen content. Turbidity shuts out sunlight, and green plants disappear. The water appears gray. Progressive reduction in living forms takes place and bacteria flourish. Sludge deposits appear on the bottom. Greasy, mosslike gray coatings of *Sphaerotilus natans* appear on the stones. The zone limit is reached when the dissolved oxygen is reduced to 40 percent of saturation, a limit set at a time when it was believed that 40 percent represented the critical oxygen situation for fish.

2. *The zone of active decomposition*   In this zone the oxygen decline continues, and may reach zero. Aquatic life may be reduced to anaerobic

bacteria and a few species of small protozoa, *Bodonidae,* which can survive without oxygen. The water appears black. Actively decomposing sludge blankets rise from the bottom, buoyed up by gases of decomposition. The water has a foul odor. In some reaches the bottom in the shallows and in the lee of stones becomes covered with tubificid worms and

Figure 18-1 *Oxygen saturation deficit in a polluted stream.*

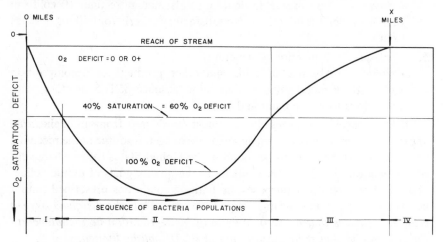

I. Degradation: Aerobic decomposition of organic matter.

II. Active Decomposition: Anaerobic decomposition of organic matter.

III. Recovery: Aerobic stabilization of products of anaerobic decomposition.

IV. Cleaner Water: Aquatic growth due to fertilizer effect of stabilized wastes.

other pollution indicators. Eventually the cycle of decomposition reaches the point where the BOD is exerted slowly (second stage, see Chapter 2) and the natural processes of reaeration overcome those of deoxygenation. D.O. again appears, and at 40 percent saturation the arbitrary limit of the zone is reached.

3. *The zone of recovery*   Aquatic plants and other aquatic life begin to reappear as the zone progresses. The sewage characteristics of the stream disappear and oxygen recovery continues. Nitrogen compounds may include ammonia and some nitrates. The zone is considered to end when the normal D.O. content of the stream is restored. Its terminus is not rigidly defined.

4. *The zone of cleaner water*   In this zone the water is back to its original dissolved oxygen content. Aquatic life flourishes. Bacteria remain high and may include pathogens. The dissolved solids are higher, espe-

cially nitrates, chlorides, sulfates, etc., but the stream is generally suitable for reuse, especially if tributaries have added to the dilution factor.

If the amount of sewage is not great the oxygen may merely sag without endangering fish life, and the added nutrient of sewage may merely fertilize the stream, making it beneficial rather than detrimental to aquatic life. If other uses of the stream are not impaired, it might be considered that the stream is doing cheaply a part of the job of sewage treatment. This is often the case when primary treatment alone is sufficient, or where final effluents are discharged in relatively large amounts.

Throughout the sequence of events depicted in Figure 18-1, processes of reaeration are active. These include:

1. Diffusion of oxygen downward at a rate proportional to the degree of undersaturation
2. Absorption of oxygen by undersaturated water broken up or distorted as it flows over rapids, around stones, etc.
3. Growth of algae and other aquatic plants in the water

The low point on the oxygen saturation-deficit curve of Figure 18-1 represents the point where depletion of substrate enables the processes of re-aeration to supply oxygen faster than the remaining BOD is exerted, and so indicates the point at which the stream begins to recover its oxygen resource.

THE OXYGEN SAG CURVE

The curve depicted in Figure 18-1 is commonly called an oxygen sag curve [2] and may be developed mathematically from the first-stage BOD curve (Figure 2-4) of the waste, and a curve representative of the reaeration rate of the receiving water in a manner briefly summarized later. It is subject to numerous limitations but is nevertheless an attractive device for estimating the effect on the oxygen resources of a stream of discharging sewage effluent of known BOD and dissolved oxygen characteristics. For this reason its development, use, and limitations are summarized even though nothing new is added to well-known and widely disseminated knowledge.

SUMMARY OF DEVELOPMENT
OF THE OXYGEN SAG CURVE

From theoretical and experimental data it has been shown that depletion of the food supply which governs the rate of oxygen demand (BOD) during the first stage of aerobic degradation (see Figure 2-5) is a first-order reaction in which the reaction rate is proportional to the concentration of the food supply at any time. This may be applied to the first-stage BOD curve as follows, to develop an equation (1) which enables the engineer to estimate the BOD at 20°C for any time t within the first stage, when oxygen demand is most significant.

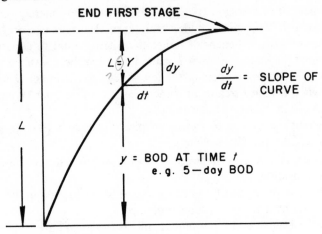

Where L = ultimate first-stage BOD
   y = BOD exerted in time t
   L_t = O_2 demand in time t
   k = reaction constant, base e
   k' = reaction constant, base 10

$\dfrac{dy}{dt}$ = increase in BOD per unit of time t

Increase also equals magnitude $\times$ rate = $(L - y)k$, or

$$\frac{dy}{dt} = (L - y)\,k$$

which integrates to form [3]

$$y = L(1 - e^{-kt}) \quad \text{or} \quad y = L(1 - 10^{-k't})$$

A more convenient form of equation results from transforming the equation and taking the logarithm of both sides.

$$\frac{L - y}{L} = 10^{-k't} \qquad \text{therefore} \qquad \log\frac{L - y}{L} = -k't \qquad (18\text{-}1)$$

Since both L and k' are unknown, they must be determined experimentally by observing the oxygen depletion in a series of replicate samples at 1, 2, 3, ..., n days at 20°C. From such experiments the commonly used value of $k' = 0.10$ has been determined for domestic sewage. For other organic wastes experimental work may be necessary.

It remains then only to obtain equations for k' and L at temperatures other than 20°C in order to make possible the development of the oxygen depletion curve resulting from the discharge of sewage, as in Figure 18-1. Such equations, developed by Streeter and Phelps [4], and Theriault [5], respectively, are as follows:

$$k'_{T°C} = k'_{20°C}(1.047^{T-20})$$ (18-2)

$$L_{T°C} = L_{20°C}[1 + 0.02(T - 20)]$$ (18-3)

From a consideration of gas transfer rate in relation to oxygen saturation deficit it can be shown that the deficit at any time t is a function of initial oxygen concentration, time, and velocity rate of transfer k_2; that is,

$$d_t = d_0 e^{-k_2 t} \quad \text{or} \quad d_t = d_0 10^{-k'_2 t}$$ (18-4)

Also:

$$k_2 \text{ at } T = k'_2 \text{ at } 20°C \ \theta^{T-20}$$ (18-5)

where θ has been evaluated by Becker [6] as 1.0159 for a river.

The experimental work required to evaluate k'_2 is so long and tedious that only a few determinations have been made; for example,

1. Mean value for Ohio River $= 0.24 = k'_2$ [4]
2. Mean value for Illinois River $= 2.57$ (turbulent condition in cold weather) [4]
3. Quiescent water $= 0.115$ [7]
4. Small ponds and backwaters $= 0.05 - 0.10$ [8]
5. Sluggish streams and large lakes $= 0.10 - 0.15$ [8]
6. Large streams, low velocity $= 0.15 - 0.20$ [8]
7. Large streams, normal velocity $= 0.20 - 0.30$.[8]
8. Swift streams $= 0.30 - 0.50$ [8]
9. Rapids and waterfalls $= 0.50$ and higher [8]

Integrating Eqs. (18-1) with (18-4) results in the classic Streeter and Phelps [4] equation which is the oxygen sag curve shown in Figure 18-1.

$$D_t = \frac{k'L}{k'_2 - k'}(10^{-k't} - 10^{-k'_2 t}) + D_a(10^{-k'_2 t})$$ (18-6)

in which D_a is the initial oxygen saturation deficit and other factors are as previously noted.

USE OF
OXYGEN SAG EQUATION

The oxygen sag equation has often been employed to make estimates of the type suggested in the example that follows.

Example   It is proposed to discharge 2 mgd of sewage into a clean stream whose minimum expected flow is 40 cfs. The 5-day 20°C BOD of the sewage is 200 ppm and the sewage is sufficiently fresh to contain 6 ppm D.O. at the time it enters the stream. The 5-day BOD of the stream is 1 ppm and it is saturated with oxygen. Previous observations of the stream indicate that at times of low flows it has a velocity of 1 ft/sec (average over several miles) and that its temperature will be a maximum of 25°C. Assuming that the temperature of the sewage is also 25°C, and that the values of k' and k'_2 at 20°C are 0.10 and 0.24, respectively, calculate:

(a)   Length of zone of degradation in miles
(b)   Length of zone of active decomposition in miles
(c)   Miles of stream unsuitable for aquatic life by Ellis's criterion (5 ppm)

Solution $Dt = \dfrac{k'L}{k'_2 - k'}(10^{-k't} - 10^{-k'_2 t}) + D_a(10^{-k'_2 t})$

D.O. saturation at 25°C = 8.38 ppm   [3]

$k'_{(25)} = 0.10 \times 1.047^{25-20} = 0.126$

$k'_{2(25)} = 0.24 \times 1.016^{25-20} = 0.26$

D.O. of mixture of water and sewage:

$$\frac{(8.38 \times 40) + (6 \times 3.1)}{43.1} = 8.2 \text{ ppm}$$

Initial D.O. saturation deficit $D_a = 8.38 - 8.20 = 0.18$ mg/liter
BOD of mixture of water and sewage:

$$\frac{(40 \times 1) + (3.1 \times 200)}{43.1} = 15.3 \text{ mg/liter}$$

First-stage 20°C BOD $= L$; $\dfrac{L - 15.3}{L} = 10^{-0.10 \times 5} = \dfrac{15.3}{0.68}$ mg/liter

$25°C$ BOD $= L_{25} = \dfrac{15.3}{0.68}[1 + 0.02(25-20)] = 24.75$ mg/liter

$D_t = \dfrac{0.126 \times 24.75}{0.26 - 0.126}[10^{-0.126t} - 10^{-0.26t}] + 0.18 \times 10^{-0.26t}$

Values of D_t and t are computed and plotted as in Figure 18-2. From this figure and using the velocity value of 1 ft/sec:

(a)   Zone of degradation = 1.2 days, or 19.6 miles
(b)   Zone of active decomposition = 3 days, or 49 miles
(c)   Unsuitable for fish life = 5.5 days, or 90 miles

**LIMITATIONS OF
OXYGEN SAG CURVE**

Computations such as those illustrated in the example can be applied to much more complicated situations in which a series of discharges and unpolluted inputs reach the stream along its course. It is only necessary to recompute the D.O. and BOD of mixture values at each change in flow and begin a new curve at the point where the change occurs. The major difficulty is that all such computations are naïve at best because of many limitations of the method illustrated.

Figure 18-2 *Oxygen sag curve.*

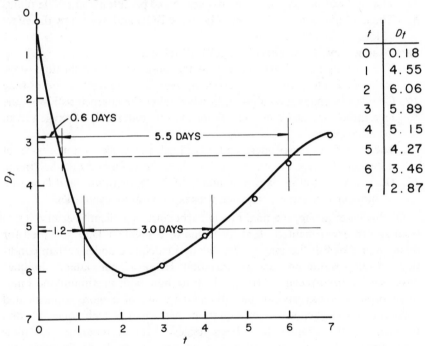

t	D_t
0	0.18
1	4.55
2	6.06
3	5.89
4	5.15
5	4.27
6	3.46
7	2.87

NOTE: ZONE OF DEGRADATION DEFINED BY 40% D.O. SATURATION, HENCE BY 60% SATURATION DEFICIT
D_t = 8.38 x .60 = 5.05 ppm
ALSO, 5 ppm D.O. = 3.38 ppm DEFICIT

Attention has already been called to the uncertainty of the assumed value of k_2 for any particular case. Dobbins [9] has listed other factors which are of considerable importance:

1. The removal of BOD by sedimentation or adsorption
2. The addition of BOD along the stretch by the scour of bottom deposits or by the diffusion of partly decomposed organic products from the benthal layer into the water above
3. The addition of BOD along the stretch by local runoff
4. The removal of oxygen from the water by diffusion into the benthal layer to satisfy the oxygen demand in the aerobic zone of this layer
5. The removal of oxygen from the water by purging action of gases rising from the benthal layer
6. The addition of oxygen by the photosynthetic action of plankton and fixed plants
7. The removal of oxygen by the respiration of plankton and sessile plants
8. The continuous redistribution of both the BOD and oxygen by the effect of longitudinal dispersion

Others added by Hansen and Frankel [10] include:

9. The variation of k', particularly at the onset of the nitrification stage which precludes assuming k' constant over any extended period of time
10. Changes in channel configuration which alter the characteristics of surface turbulence and consequently of the rate of transfer of oxygen from the atmosphere
11. The effects of suspended and dissolved substances on the rate of diffusion of oxygen from the surface into the main body of the stream
12. Diurnal variation in oxygen content, BOD, temperature, and flow rate of influent discharges, whether wastes or natural tributaries

Of this latter group, the final one is particularly significant because it has been shown experimentally that the value of 5 mg/liter D.O. assumed for maintenance of fish life may be drastically inadequate under certain conditions of temperature and diurnal variation. However, computer techniques have been utilized recently [10] to illustrate how both maximum and minimum profiles of oxygen can be estimated by use of a more sophisticated version of the oxygen-sag-curve equation. A number of techniques, beyond the scope of this chapter, have been suggested for overcoming the limitations cited above. Of particular interest are proposals of Dobbins [9], Camp [11], Li [12], O'Connor [13], Thomann [14], Velz [15, 16], and Thomas [17].

EVALUATION
OF "DILUTION" METHOD
OF WASTE TREATMENT

As noted previously, the necessity for returning domestic and industrial flows to the freshwater resource, together with economic and technical limitations of conventional purification processes, makes inevitable the imposition of some of the water quality management load upon the resource. On the other hand, the rising tide of "pure water" objectives of society is opposed to the concept of waste treatment as an acceptable beneficial use of water. To reconcile these two opposing factors it seems certain that the parameters of quality change in flowing water will have to be refined further so that the effects on water quality of utilizing the treatment potential of streams will be subject to more precise estimation.

What began as "sewage disposal by dilution" must now become a treatment method subject to scientific and engineering control, even though it is relegated from a primary to a tertiary method of waste treatment. Specifically, it seems unlikely that mere sequestering of wastes within the water mass, which characterizes dilution as a disposal method, will continue to be acceptable in the case of domestic return flows. Thus in most cases the role of the stream in water quality management will be that of an acceptor of effluents from other engineered systems at a level of BOD compatible with the predictable capacity of the stream to handle such oxygen demand. On the other hand, perfection of conventional treatment systems can at best only lead to the discharge of stabilized compounds. These the stream must accept and dilute, certainly until such time as quality management may require partial deionization of return flows. In the case of industrial wastes which are not organic in nature, dilution will continue to be a function of the receiving water.

Disposition of these quality factors by the stream may involve a combination of simple transport to the ocean, precipitation and later pickup by floodwaters, support of a more abundant aquatic society, increased mineralization, and a variety of more subtle equilibria.

There is no way that the flowing stream can be eliminated as a waste-water treatment system involved in attaining the goals of water quality management.

REFERENCES

1   M. M. Ellis: *Detection and Measurement of Stream Pollution*, U.S. Bureau of Fisheries Bulletin 22, 1937.

2   H. A. Thomas, Jr.: "Slope Method of Evaluating the Constants of the First-Stage BOD Curve," *Sewage Works J.*, 9(4), 1937.

3   Gordon M. Fair and J. C. Geyer: *Water Supply and Waste Disposal*, John Wiley & Sons, Inc., New York, 1954.

4   H. W. Streeter and E. B. Phelps: *A Study of the Pollution and Natural Purification of the Ohio River*, U.S. Public Health Bulletin 146, 1925.

5   E. J. Theriault: "The Oxygen Demand of Polluted Waters," U.S. Public Health Bulletin 173.

6   H. G. Becker: "Mechanism of Absorption of Moderately Soluble Gases in Water," *Ind. Eng. Chem.*, 16, 1924.

7   W. E. Adeney: "The Principles and Practice of the Dilution Method of Sewage Disposal," Cambridge University Press, London, 1928.

8   *Report of the Board of Review, Sanitary District of Chicago*, Part III, Appendix I, February 21, 1925.

9   W. E. Dobbins: "BOD and Oxygen Relationships in Streams," Proc. Paper 3949, *J. Sanit. Eng. Div., ASCE*, 90(SA3):53, June, 1964.

10   W. W. Hansen and R. J. Frankel: "Economic Evaluation of Water Quality: A Mathematical Model of Dissolved Oxygen Concentration in Freshwater Streams," Second Annual Report, Sanitary Engineering Research Laboratory, University of California, SERL Rept. 64-11, August, 1965.

11   T. R. Camp: *Water and Its Impurities*, Reinhold Publishing Corporation, New York, 1963.

12   W. Li: "Unsteady Dissolved-oxygen Sag in a Stream." Proc. Paper 3129, *J. Sanit. Eng. Div., ASCE*, 88(SA3):75, May, 1962.

13   D. J. O'Connor: "The Effect of Stream Flow on Waste Assimilation Capacity," *Proc. Seventeenth Purdue Industrial Waste Conf.*, Purdue University, May, 1962.

14   R. V. Thomann: "Mathematical Model for Dissolved Oxygen," Proc. Paper 3680, *J. Sanit. Eng. Div., ASCE*, 89(SA5):1, October, 1963.

15   C. J. Velz and J. J. Gannon: "Biological Extraction and Accumulation in Stream Self-purification," Publication, School of Public Health, University of Michigan, Ann Arbor.

16   C. J. Velz: "Significance of Organic Sludge Deposits," *Oxygen Relationships in Streams, Proc. Seminar, Water Supply and Water Pollution Program*, Robert A. Taft Sanitary Engineering Center, U.S. Public Health Service, March 1958.

17   H. A. Thomas, Jr.: "Pollution Load Capacity of Streams," *Water and Sewage Works*, 95, 1948.

CHAPTER 19

Marine and estuarine disposal of wastes*

INTRODUCTION

Inevitably the ocean must be regarded as the ultimate sink into which all water quality factors flow and in which they are left behind in the hydrological cycle. This does not mean, however, that there is no need of concern for the type of wastes the ocean may receive once such wastes have survived the considerations of beneficial use, preservation of resources, and sentimentality which accompany quality management in the freshwater system. Beyond the limits of even the estuarine resources discussed in Chapter 7 lie resources of the ocean which grow in importance as human population of the land mass increases. Radioactivity and pesticides, for example, are among the quality factors which may be concentrated in marine life and spread rapidly over the world's oceans by marine organisms with results not yet fully evaluated by ecologists.

In general, however, it may be assumed that the upstream needs of mankind will engender the engineered systems which will effectively prevent water quality considerations of the ocean from becoming the controlling factor in freshwater quality management. Nevertheless, great population concentrations exist at the shoreline, and the return flows from their domestic and industrial use of water must either be reclaimed or discharged into marine waters. In either case, most of the quality factors reach the ocean in either a raw or a stabilized condition.

Life on the seacoast likewise generally involves shipping and other ocean transport activities. This historically has created cities where bays and estuaries provide harbor facilities. Consequently, the temptation, and often the necessity, exists for discharge of waste waters into the estuarine waters, or into the shallow ocean waters overlaying the continental shelf.

*This chapter authored by R. E. Selleck.

If this is to be accomplished in a manner acceptable to human aesthetics, and without interference with man's use of other shoreline resource values, it may be necessary to precondition waste waters by the engineered systems discussed in other chapters, i.e., ponds, soil systems, conventional works, etc. In other cases, direct discharge of raw waste waters may be feasible. In any event, an engineered system of discharge is required, the general principles of which are discussed in this book.

FACTORS OF SIGNIFICANCE
IN AN ENGINEERED SYSTEM

Under steady conditions the concentration of a waste-water constituent, or quality factor, in a receiving water will usually be maximum at the point of discharge. The magnitude of this maximum concentration depends upon the rate of transport of the discharged mass away from the outfall. This transport rate is determined by such factors as the current and turbulent dispersion characteristics of the system, the interactions of the quality factor of concern with the receiving water, and the type of outfall installation utilized. Generally, rather gross approximations have to be made before these factors can be formulated mathematically. Some of the approximations often utilized in practice are described in the following sections.

MAXIMUM CONCENTRATION IN THE
PRESENCE OF APPRECIABLE CURRENT

Rapid initial mixing is normally desired to prevent excessive local concentrations at the point of waste-water discharge. To this end it is customary to assume that waste water discharged from a line source is instantly mixed in the water volume element overlying the line source. In practice a line source is equivalent to a diffuser in which the waste water is released through several open ports closely spaced along a submerged pipeline. The line source assumption is often used in the following situations:

1. Outfalls, with or without diffusers, in vertically well-mixed estuaries if the waste discharge is relatively great in comparison to the boundaries of the estuary

2. Marine (coastal) outfalls having a multiport diffuser at the discharge end

3. Stream outfalls under conditions similar to (*1*), above.

When the advective current over the outfall is large compared to the transport effected by turbulent dispersion, and the rate of decay, degradation, or die-away of the quality factor to be controlled is relatively slight, it is often assumed that advection is the sole means of mass transport in the region. In this case the waste-water release is assumed to be a line source placed normal to the direction of current flow as illustrated in Figure 19-1. The mass of the quality factor released by the outfall per unit time δt is $Q_w c_w \delta t$, where Q_w is the rate of discharge and c_w is the waste-water concentration of the constituent of concern. The mixing element overlying the outfall has an area A and length $U\delta t$, where U is the advective velocity, as designated in Figure 19-1.

Figure 19-1 *Mixing element overlying line source and initial concentration of water quality factor—advection only.*

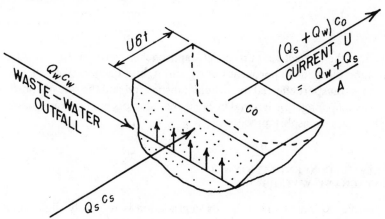

The diluting water flowing into the mixing element has a flow rate of Q_s and a background concentration c_s of the constituent. Consequently, the background mass of the constituent transported into the element is $Q_s c_s \delta t$.

From mass continuity considerations it is apparent that

$$c_o = \frac{Q_s c_s \delta t + Q_w c_w \delta t}{A U \delta t} = \frac{Q_s c_s + Q_w c_w}{Q_s + Q_w}$$

where c_o is the concentration of the quality factor within the mixing element. The above relationship may be written in the form

$$S = \frac{Q_w + Q_s}{Q_w} = \frac{c_w - c_s}{c_o - c_s} \tag{19-1}$$

where S is the hydraulic dilution.

In the particular case where $c_s \to 0$, the above equation reduces to

$$S = \frac{c_w}{c_o} = 1 + \frac{Q_s}{Q_w} \tag{19-2}$$

And if $Q_s >> Q_w$, then

$$\frac{c_o}{c_w} \cong \frac{Q_w}{Q_s} \tag{19-3}$$

Equation (19-3) is often used in the design of multiport marine outfalls which often approximate line sources. Letting

$$Q_s = Ubh$$

where U is the design speed of the current, b the length of the line source (normal to the current), and h the effective depth of the resulting field of waste water mixed with the ocean water, (19-3) reduces to

$$\frac{c_o}{c_w} \cong \frac{Q_w}{Ubh} \tag{19-4}$$

When the advective current Q_s approaches zero in Eq. (19-1), the hydraulic dilution S approaches unity and c_o approaches c_w. In such a case the mass transport effected by turbulent dispersion and the rate of constituent decay can no longer be ignored because these factors are now rate controlling in the mass transport process.

MAXIMUM CONCENTRATION IN STAGNANT WATERS

If the rate of constituent decay or degradation is relatively slight, as is the case with many chemical compounds, and the horizontal dimensions of the receiving body of water are not limiting in the sense of mass transport, the dilution and dispersion observed in rising jets may be used to estimate the concentration of a waste constituent over the outfall. The factors which determine the dilution effected in this manner are the momentum of the fluid at the discharge port, the buoyancy of the waste water in the receiving water, the angle of the jet relative to the water surface, the depth and the degree of stratification of the receiving water above the jet.

A vertical jet having pure momentum, i.e., no buoyancy in a homogeneous receiving water, is the simplest case to treat mathematically. Abraham [1] and Hart [2] show that the hydraulic dilution S effected within the core of the rising jet at an elevation y above a circular port of diameter D is equal to 0.192 y/D (when $y/D \geqslant 5.2$). It is apparent, however, that in a system having a finite depth of water over the port the free water surface acts as a physical boundary. Some of the constituent mass is reflected back downward into the rising jet at this boundary and a surface boil is created.

The dilution and dispersion occurring in the boil are not well understood. However, it has been observed by Frankel and Cumming [3] that a large degree of turbulence exists in the boil and it may be that the boil is more or less homogeneous in composition. Also, the influence of the boil appears to extend rather deeply along the axis of the jet. Frankel and Cumming estimate the depth of influence to be about 0.25 H/D, where H is the total water depth over the port.* In general it is the dilution S_o obtained in the boil which is of concern to the designer of waste-water outfalls attempting to meet water quality criteria. In the case of the vertical jet having pure momentum the magnitude of S_o should be about 0.144 H/D if the boil extends to a depth of about 0.25 H/D. The above relationships are illustrated in Figure 19-2.

*The penetration depth of the boil should not be confused with the effective depth of mixing of the waste water in the receiving water stipulated in Eq. (19-4). The two depths represent different phenomena and are not necessarily related.

Figure 19-2 *Dilution effected in a vertical, pure momentum jet* ($\mathbf{F} = \infty$).

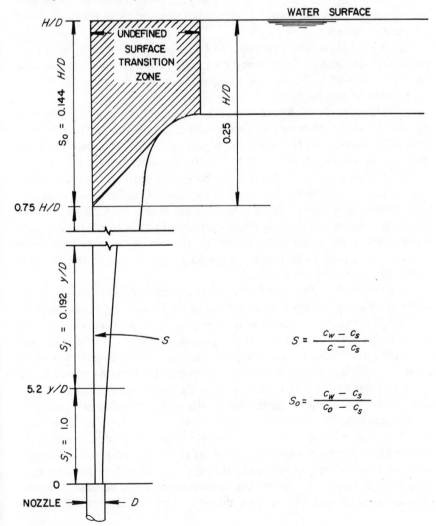

$$S = \frac{c_W - c_S}{c - c_S}$$

$$S_o = \frac{c_W - c_S}{c_o - c_S}$$

Buoyancy is created when the density of the jetted fluid is less than that of the surrounding water. Relative buoyancy is often measured by the Froude number

$$F = \frac{V}{\sqrt{gD\,[(\rho_s - \rho_w)/\rho_w]}}$$ (19-5)

where V is the mean jet velocity at the port opening, ρ_s the density of the receiving water, and ρ_w the density of the waste water. The value of the Froude number decreases as the difference between the densities of the receiving and waste water increases. (In the case of a pure momentum jet the difference in densities is zero and $F = \infty$.) The observed value of F usually ranges from 3 to about 20 for a typical domestic waste water released into marine waters.

Figure 19-3 shows observed values of S_o reported by Rawn, Bowerman, and Brooks [4] and by Frankel and Cumming [3] for various Froude numbers and ratios of H/D. Two port angles, horizontal and vertical, are also represented in the figure.

For vertical ports the magnitude of S_o appears to increase at an increasing rate as the jet buoyancy increases (decreasing F number). The lower limit of S_o appears to be within the region of the pure momentum jet as predicted by the equation $S_o = 0.144\,H/D$ described previously.

On the other hand, the value of S_o for horizontal ports appears to decrease at a decreasing rate as the buoyancy of the jetted fluid increases, with relatively little change being effected for F values of less than 20. One reason for the greatest dilution being effected at high Froude numbers with horizontal ports is the fact that as F approaches infinity the only mechanism available for transport of constituent mass to the surface is the turbulent dispersion created by the jet itself.

Discussion of jet dilution has thus far been restricted to receiving waters which are homogeneous from the surface to the bottom. Practical situations are often encountered where the receiving water is stratified into two (or more) layers with the denser layer lying along the bottom. The performance of a fluid jetted in such waters depends on the residual momentum of the jet at the interface of the layers and the degree of dilution effected in the rising jet prior to reaching the interface. If the momentum of the jet is sufficiently great and/or the dilution with the denser water sufficiently small, the jet will penetrate the interface and rise to the water surface, creating a surface boil. The waste-receiving water mixture may then remain within the surface layer or plunge back into the lower layer, depending on the density of the mixture in the boil relative to that of the surface layer. In some cases the jet will not penetrate the interface between the layers because of insufficient momentum and/or excessive dilution in the lower layer. The waste effluent will then be confined within the lower layer—a situation analogous to an atmospheric inversion layer of sufficient strength to confine smoke plumes within a smogbound basin.

It is generally considered desirable in cases of outfall design to maintain the waste effluent in the lower, more dense layer of stratified waters. In this way no evidence of the waste effluent, aside from flotables, will generally be visible at the surface. The concept, however, involves no concern for the

Figure 19-3 *Dilution of rising jets effected at water surface for homogeneous receiving water.*

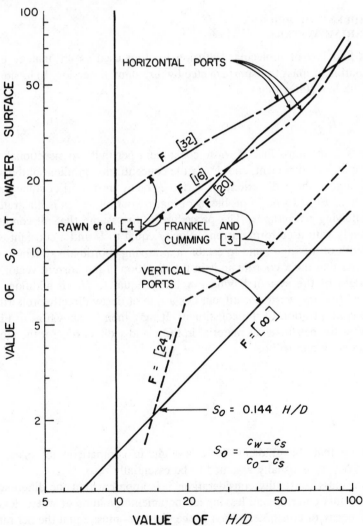

ecology of the lower stratum of water and hence may not represent adequate engineering for many situations in the future. Severe stratification, however, is often a temporary phenomenon in many natural bodies of water. Unless the stratification is known to be permanent, waste-water outfall structures are generally designed on the basis of the waste effluent reaching the surface and being maintained subsequently in the surface layer.

The factors which determine the performance of a jet in stratified waters are involved. For this reason no quantitative description of the phenomenon is presented herein. The interested reader is referred to the discussions of Hart [2] and Frankel [5].

TURBULENT DISPERSION IN MARINE WATERS

Fick's first law of molecular diffusion is often used as an analogy in the formulation of mass transport effected by turbulent dispersion. In its simplest form the law states that

$$N = -E\frac{dc}{dx} \tag{19-6}$$

where N is the mass flux of waste constituent per unit cross-sectional area, E a dispersion coefficient, and dc/dx the concentration gradient of the waste constituent in the x direction, all variables being measured at some time t. The use of Eq. (19-6) in problems of marine outfall design is illustrated in the following paragraphs. The procedure is to assume that all conditions are steady with a uniform current velocity equal to U and a constant effective depth of mixing h, and to establish a moving coordinate system in such a manner that the x coordinate is in the direction of the current vector with the origin of the system having a velocity equal to U. In addition, it is assumed that the waste constituent has a rate of decay directly proportional to the concentration of the constituent. It may then be shown as in Figure 19-4 that the net fluxes of material in the x and y directions for the moving coordinate system will be

$$N_x = -E_x\frac{\partial c}{\partial x}$$

$$N_y = -E_y\frac{\partial c}{\partial y}$$

Assuming that the gradient $\partial c/\partial x$ is slight in comparison to $\partial c/\partial y$, then $N_x << N_y$. N_x is usually assumed to be essentially zero.

From mass continuity considerations it is apparent that the difference in the mass flux entering and leaving an incremental volume of water—less the rate of decay of constituent mass in the element—must equal the net rate of change of constituent mass in the element, or

$$-\frac{\partial N_y}{\partial y} - kc = \frac{\partial c}{\partial t}$$

where k is the rate of constituent decay. The above equation reduces to

$$\frac{\partial}{\partial y}\left(E_y \frac{\partial c}{\partial y}\right) - kc = \frac{\partial c}{\partial t} \tag{19-7}$$

upon the substitution of Fick's first law [Eq. (19-6)].

Since $x = Ut$, Eq. (19-7) also can take the form

$$\frac{\partial}{\partial y}\left(E_y \frac{\partial c}{\partial y}\right) - kc = U \frac{\partial c}{\partial x} \tag{19-8}$$

when the coordinate system is considered fixed in space.

Field observations indicate that the value of the coefficient E_y increases with the scale of the spread of the waste-water plume. Pearson [6], as well

Figure 19-4 *Mass flux through an elemental volume of marine water.*

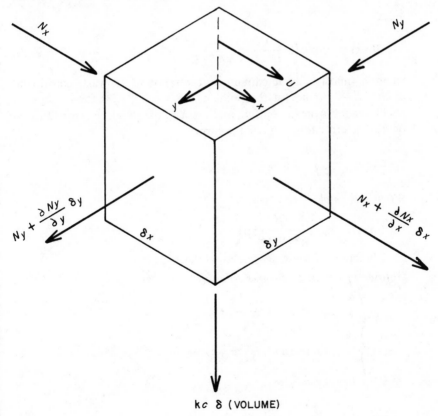

$kc\ \delta\ (\text{VOLUME})$

as other investigators, has observed that the coefficient appears to be approximately proportional to the ⅘ power of the scale of the field dispersion. Brooks [7] has proposed the equation

$$E_y = \alpha (2 \sqrt{3} \, \sigma_y)^{4/3} \qquad\qquad (19\text{-}9)$$

for evaluating E_y, where σ_y is the standard deviation of the distribution of the waste constituent in the direction normal to the axis of the plume. The coefficient α appears to have a value of about 0.001 when σ_y is measured in feet and E_y is in ft²/sec.

Two example solutions of Eq. (19-8) follow.

1.   Constant E_y

   Diffuser of length b

$$c = \begin{cases} c_o & \text{at } x = 0 \text{ and } -\dfrac{b}{2} \le y \le \dfrac{b}{2} \\[2mm] 0 & \text{at } x = 0 \text{ and } -\dfrac{b}{2} > y > \dfrac{b}{2} \\[2mm] 0 & \text{at } y = \pm\infty \end{cases}$$

Then

$$c(\max) = c_o \exp\left(\frac{-kx}{U}\right) \operatorname{erf}\left(\frac{b}{4}\sqrt{\frac{U}{E_y x}}\right) \qquad\qquad (19\text{-}10)$$

where $c\,(\max)$ is the maximum concentration of the waste constituent in the waste plume (at the point where $y = 0$), c_o is the constituent concentration over the outfall, and erf X is the standard error function of X. The error function has the form

$$\operatorname{erf} X = \frac{2}{\sqrt{\pi}} \int_0^X e^{-j^2} \, dj$$

When b is small

$$c(\max) \to c_o b \sqrt{\frac{U}{4\pi E_y x}} \exp\left(\frac{-kx}{U}\right) \qquad\qquad (19\text{-}11)$$

2. E_y a function of the field spread to the ⅘ power

   Diffuser of length b (Brooks solution [7])

$$c = \begin{cases} c_o & \text{at } x = 0 \text{ and } \dfrac{-b}{2} \le y \le \dfrac{b}{2} \\[2mm] 0 & \text{at } x = 0 \text{ and } \dfrac{-b}{2} > y > \dfrac{b}{2} \\[2mm] 0 & \text{at } y = \pm\infty \end{cases}$$

$E_y = E_y'$ at $x = 0$

E_y determined by Eq. (19-9)

Then

$$c\,(\text{max}) = c_o \exp\left(\frac{-kx}{U}\right) \text{erf}\left[\frac{3/2}{\left(1 + \frac{2}{3}\beta\frac{x}{b}\right)^3 - 1}\right]^{\frac{1}{2}} \qquad (19\text{-}12)$$

where $\beta = 12\ E_y'/Ub$, with the remaining terms being equivalent to those given previously.

The value of c_o used in the solutions given above for comparatively large values of b is generally determined from Eq. (19-4), i.e.,

$$c_o \cong \frac{Q_w c_w}{Ubh}$$

In well-stratified waters the depth of effective mixing h may be taken equal to the average depth of the layer in which the waste-water field is confined. In poorly stratified waters, however, the buoyant waste effluent will generally first rise to the surface and then gradually diffuse vertically downward throughout the entire water column as the waste mass is transported away from the outfall. In that case h will not be constant, and a two-dimensional differential equation in terms of both horizontal and vertical spread is required to obtain an accurate solution. Approaches of this type are described by Foxworthy, Tibby, and Barsom [8].

USE OF EXAMPLE
SOLUTION TO EQ. (19-8)

Example   The critical current speed over an ocean outfall having an effective diffuser length of 1,000 ft is 0.5 knot (0.85 fps). The effective depth of mixing is 30 ft. Using Eqs. (19-4) and (19-10), compute the maximum concentration of coliform organisms at a distance of 2 nautical miles from the outfall if the waste effluent has a concentration of 10^6 organisms per milliliter and a flow of 100 cfs. The rate of decay of the organisms in the receiving water is 0.5 per hour, and the average coefficient of lateral dispersion is estimated to be 50 ft²/sec.

Solution $c_o \cong \dfrac{c_w Q_w}{Ubh}$ $\qquad\qquad\qquad\qquad\qquad\qquad$ (19-4)

$$c_o \cong \frac{10^6 \times 100}{0.85 \times 1000 \times 30} = 3.92 \times 10^3\,\text{ml}^{-1}$$

$$c\,(\text{max}) = c_o \exp\left(-\frac{kx}{U}\right) \text{erf}\left(\frac{b}{4}\sqrt{\frac{U}{E_y x}}\right) \qquad (19\text{-}10)$$

$$\frac{kx}{U} = \frac{0.5 \times 2}{0.5} = 2$$

$$X = \frac{b}{4}\sqrt{\frac{U}{E_y x}} = \frac{1000}{4}\sqrt{\frac{0.5}{50 \times 3600 \times 2}} = 0.294$$

$$\text{erf } X = 0.323$$

$$c\,(\text{max}) \cong \frac{3.92 \times 10^3 \times 0.323}{7.40} = 170 \text{ organisms per milliliter}$$

TURBULENT DISPERSION OF ESTUARINE WATERS

The solutions of Eq. (19-8) presented above are generally applicable to the design of marine outfalls if one has some information regarding the design values of the depth of effective mixing as well as the direction and speed of the advective current. These solutions, however, usually cannot be used in estuarine problems because the water mass is bounded by physical barriers having finite but unknown values of constituent concentration. An approach often used in problems of this kind is to section the estuary into finite elements which are constructed normal to the longitudinal axis of the estuary as illustrated in Figure 19-5. It is then assumed that the value of a waste-effluent constituent concentration at some longitudinal distance x

Figure 19-5   *Mass flux through an elemental estuarine segment.*

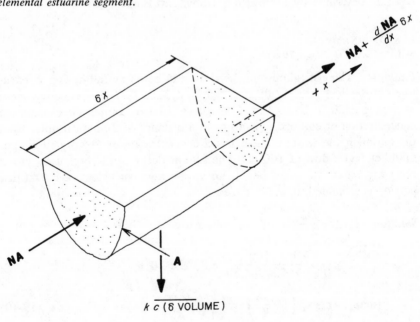

$$k\,\overline{c}\,(\delta\,\text{VOLUME})$$

in the estuary represents an average value taken over the entire estuarine cross section and for a period of time equal to at least one tidal cycle. The coefficient of longitudinal dispersion E_x is also assumed to represent the mean value taken over a complete tidal cycle. Using these assumptions, Eq. (19-8) may be derived in the form

$$\frac{d}{dx}\left(\mathbf{A}\mathbf{E}_x\frac{d\mathbf{c}}{dx}\right) - \frac{d}{dx}(\mathbf{A}\mathbf{U}\mathbf{c}) - \mathbf{A}k\mathbf{c} = 0 \qquad (19\text{-}13)$$

using the principle of mass continuity. The boldface symbols indicate that the values are cross-sectional averages taken over a complete tidal cycle. The coefficient $\mathbf{U}$ represents the mean net advective velocity remaining in the estuary after the tidal velocities have been discounted, and $\mathbf{A}$ the mean cross-sectional area of the estuary. The values of $\mathbf{U}$, $\mathbf{A}$, and $\mathbf{E}_x$ may change with location so it is necessary to include those coefficients inside the differentiated members of Eq. (19-13).

As in the case of marine dispersion the coefficient $\mathbf{E}_x$ generally is not constant with x, nor can it be estimated without some prior knowledge of the mixing and dispersion characteristics of the estuary. Values of the coefficient ranging from about 500 to 10,000 ft^2/sec have been observed in natural estuaries, depending on the location of the measurement, the size and geometry of the system, the net advective velocity, and the degree of tidal energy dissipation. Usually the coefficient will be less in the landward reaches of the estuary and will increase with increasing advective velocities. In the San Francisco Bay system [9], for example, the magnitude of $\mathbf{E}_x$ appears to be at least partially a function of $(\mathbf{U}\mathbf{U}'\mathbf{H})^{3/4}$, where $\mathbf{U}$ is the mean net advective velocity, $\mathbf{U}'$ the mean tidal velocity taken without regard to sign, and $\mathbf{H}$ the average hydraulic depth. Unless the coefficients $\mathbf{E}_x$ and $\mathbf{U}$ can be expressed in relatively simple terms as functions of x, Eq. (19-13) cannot be integrated without the utilization of finite difference methods and the digital computer.

An insight on how Eq. (19-13) functions may be obtained assuming the coefficients $\mathbf{E}_x$, $\mathbf{U}$, $\mathbf{A}$, and k are independent of x. In this case Eq. (19-13) reduces to

$$\mathbf{E}_x\frac{d^2\mathbf{c}}{dx^2} - \mathbf{U}\frac{d\mathbf{c}}{dx} - k\mathbf{c} = 0 \qquad (19\text{-}14)$$

Using the boundary conditions

$$\mathbf{c} = \mathbf{c}_o \quad \text{at } x = 0 \quad \text{and} \quad \mathbf{N} = \mathbf{N}_o \quad \text{at } x = 0$$

where $\mathbf{N}$ is the average flux of constituent mass in the x direction, the solution of Eq. (19-14) is

$$\mathbf{c} = \left[\mathbf{c}_o\cosh mx + \frac{1}{m}\left(\mathbf{c}_o n - \frac{\mathbf{N}_o}{\mathbf{E}_x}\right)\sinh mx\right]\exp(nx) \qquad (19\text{-}15)$$

where

$$n = \frac{U}{2E_x} \qquad m = n \left(1 + \frac{4kE_x}{U^2} \right)^{\frac{1}{2}}$$

The value of c_o in Eq. (19-15) may not be equivalent to the type of approximation given by Eq. (19-4) and the value of AN_o may not be equal to the waste-effluent source strength $Q_w c_w$ because turbulent diffusion may play a fundamental role in the transfer of the constituent mass. If this is the case, the flux of constituent mass will be both landward and seaward of the source plane, or

$$N_s = N_e + \frac{Q_w c_w}{A} \tag{19-16}$$

where N_s and N_e are the average mass fluxes directed seaward and landward of the source plane, respectively, as shown in Figure 19-6. The value

Figure 19-6 *Example boundary conditions for constant coefficient solution to estuarine diffusion equation.*

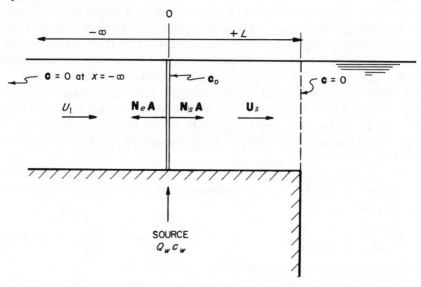

of c_o may then be estimated using Eq. (19-16) and the appropriate boundary conditions for Eq. (19-15). For example, assume that the boundary conditions are:

$$c = \begin{cases} 0 & \text{at } x = -\infty \quad \text{(landward reach)} \\ 0 & \text{at } x = L \quad \text{(seaward reach)} \end{cases}$$

Then, from Eq. (19-15),

$$\mathbf{N}_e = \mathbf{c}_o \mathbf{E}_x (n - m)_e \qquad \text{and} \qquad \mathbf{N}_s = \mathbf{c}_o \mathbf{E}_x (m \operatorname{ctnh} mL + n)_s$$

Substitution of the above in Eq. (19-16) gives

$$\mathbf{c}_o = \frac{Q_w c_w}{\mathbf{A} \mathbf{E}_x} \left[\frac{1}{(m \operatorname{ctnh} mL + n)_s + (n - m)_e} \right] \qquad (19\text{-}17)$$

where the subscripts s and e indicate the values of $\mathbf{U}$ in the seaward and landward reaches, respectively.

If by chance the waste constituent is not subject to degradation, that is, $k = 0$, then Eq. (19-17) reduces to

$$\mathbf{c}_o = \frac{2 Q_w c_w}{\mathbf{A} \mathbf{U}_s (\operatorname{ctnh} nL + 1)_s} \qquad (19\text{-}18)$$

In addition, if the value of $n_s L$ is greater than about 3.0, the equation reduces to

$$\mathbf{c}_o = \frac{Q_w c_w}{\mathbf{A} \mathbf{U}_s}$$

which is equivalent to Eq. (19-2). Thus, when the term $\mathbf{U}_s L/2\mathbf{E}_x \geq 3.0$, the effect of turbulent dispersion on the net transport of a conservative waste constituent away from a line source is negligible for the conditions specified in the derivation of Eq. (19-18).

USE OF EQ. (19-17)
IN ESTUARINE PROBLEMS

Problem The regulatory agency in a particular case requires that the average BOD_5 over an estuarine outfall cannot exceed 5 mg/liter during low tributary inflow conditions. The estuary is more or less uniform in the reach of concern and has the following characteristics:

$\mathbf{U} = 0.01$ fps (landward of source)

$\mathbf{E}_x = 500$ ft²/sec

$\mathbf{A} = 10,000$ ft²

The strength of the waste source is

$Q_w = 100$ cfs

$c_w = 200$ mg/liter

and the BOD_5 has an effective rate of decay of 2×10^{-6} per second (base e) in the receiving water. Compute the maximum distance from the seaward end of the estuary at which the outfall can be placed without exceeding the water quality criterion.

Solution $\dfrac{U}{2E_x} = n$:

$$n_e = \frac{0.01}{2 \times 500} = 1 \times 10^{-5}\,\text{ft}^{-1}$$

$$U_s = \frac{Q_t}{A} = \frac{100 + 100}{10^4} = 0.02\,\text{fps}$$

$$n_s = \frac{0.02}{2 \times 500} = 2 \times 10^{-5}\,\text{ft}^{-1}$$

$$\frac{4k\mathbf{E}_x}{\mathbf{U}^2} : \begin{cases} \dfrac{4 \times 2 \times 10^{-6} \times 500}{1 \times 10^{-4}} = 40\,\text{(landward)} \\[2ex] \dfrac{4 \times 2 \times 10^{-6} \times 500}{4 \times 10^{-4}} = 10\,\text{(seaward)} \end{cases}$$

$$m_e = 1 \times 10^{-5}\,\sqrt{41} = 6.40 \times 10^{-5}\,\text{ft}^{-1}$$

$$m_s = 2 \times 10^{-5}\,\sqrt{11} = 6.64 \times 10^{-5}\,\text{ft}^{-1}$$

From Eq. (19-17):

$$(\coth mL)_s = \frac{1}{m_s}\left[\frac{Q_w}{\mathbf{AE}_x}\left(\frac{c_w}{c_o}\right) + (m - n)_e - n_s \right]$$

$$(\coth mL)_s = \frac{10^5}{6.64}\left(\frac{100}{10^4 \times 500} \times \frac{200}{2} + 13.2 \times 10^{-5} - 2 \times 10^{-5} \right)$$

$$= 13.75$$

or

$$\left(\frac{e^{2mL} + 1}{e^{2mL} - 1} \right)_s = 13.75 \quad \text{and} \quad L = 1{,}100\,\text{ft}$$

STATUS OF ENGINEERED SYSTEMS

Marine outfalls

On the basis of such factors as discussed in the foregoing sections of this chapter, estimates have been made of current velocities, mixing depths, turbulent dispersion, bacterial concentrations, etc.; and numerous marine outfalls based on these estimates have been designed and constructed. Among the largest ever undertaken is the 12-ft-diameter submarine outfall of the City of Los Angeles, which terminates 5 miles offshore in 192 ft of water. Monitoring of the area indicates that it is successful in keeping pollutants off the beaches and transporting the mixed waste and seawater away from the area beneath the surface.

For the most part, however, the linkage between theory and design remains somewhat tenuous. Engineering judgment derived from theoretical analyses, oceanographic data, experience, and other factors cited in the literature is at present the major factor in successful marine outfall design.

Estuarine outfalls

With the advent of the modern digital computer and the evolution of systems analysis, problems of waste-water disposal in tidal estuarine waters, which until recently appeared too complex to permit evaluation in detail by any means except hydraulic (physical) models, can now be solved rather precisely and in detail using mathematical models. Such models offer great advantages over hydraulic models because complex interactions between a given waste constituent and the receiving water can be included in the mathematical model without excessive difficulty, whereas it is physically impossible to include such interactions in hydraulic models. It is now believed that the formulation of estuarine mixing and exchange has reached such a stage of development that waste-effluent concentration distributions can be computed to a greater degree of accuracy than is actually warranted by the comparatively unknown magnitudes of the reaction coefficients existing among the various parameters of a complex ecological system. It is to be expected that such methods of computation will eventually be applied to marine outfall design, but the circulation patterns of the coastal regions of the oceans are not as amenable to mathematical treatment as those found in most estuarine waters.

LIST OF SYMBOLS

A Cross-sectional area

b Length of diffuser normal to current flow

c Constituent concentration in receiving water

c_w Concentration of constituent in waste-water effluent

c_o Concentration of constituent over outfall

c_s Background concentration of constituent in receiving water

D Diameter of discharge port

E Turbulent dispersion coefficient

F Froude number

g Acceleration due to gravity

h Effective depth of mixing

H Total water depth over discharge port or hydraulic water depth

k First-order rate of decay of constituent mass

N Mass flux of constituent per unit area

Q_s Rate of advective flow
Q_w Rate of waste-water discharge
ρ Density
S Hydraulic dilution factor
S_o Hydraulic dilution in waste-water boil
U Advective flow velocity
V Jet velocity at point opening
X Longitudinal coordinate
y Distance above discharge port or lateral coordinate

REFERENCES

1   G. Abraham: "Jet Diffusion in Liquid of Greater Density," *J. Hyd. Div., ASCE*, 86(HY6):1–15, June, 1960.

2   W. E. Hart: "Jet Discharge into a Fluid with a Density Gradient," *J. Hyd. Div., ASCE*, 87(HY6):171–200, November, 1961.

3   R. J. Frankel and J. D. Cumming: "Turbulent Mixing Phenomena of Ocean Outfalls," *J. Sanit. Eng. Div., ASCE*, 91(SA2):33–59 April, 1965.

4   A. M. Rawn, F. R. Bowerman, and N. H. Brooks: "Diffusers for Disposal of Sewage in Sea Water," *J. Sanit. Eng. Div., ASCE*, 86(SA2):65–105, March, 1960.

5   R. J. Frankel: Discussion to "Tracer Studies on Jet Diffusion and Stratified Dispersion," in "Advances in Water Pollution Research," *Proc. 3d Intern. Conf. on Water Pollution Research*, Water Pollution Control Federation, Washington, D.C., 3:79, 1967.

6   E. A. Pearson: "An Investigation of the Efficiency of Submarine Outfall Disposal of Sewage and Sludge," California State Water Quality Control Board Publication 14, 1955.

7   N. H. Brooks: "Diffusion of Sewage Effluent in an Ocean Current," *Proc. 1st Intern. Conf. on Waste Disposal in the Marine Environment*, Berkeley, California, July, 1959. Pergamon Press, New York, 1960.

8   J. E. Foxworthy, R. B. Tibby, and G. M. Barsom: "Dispersion of a Surface Waste Field in the Sea," *J. Water Pollution Control Federation*, 38(7), July, 1966.

9   R. E. Selleck and B. Glenne: "A Model of Mixing and Diffusion in San Francisco Bay," vol. VII of Final Report, "A Comprehensive Study of San Francisco Bay," Sanitary Engineering Research Laboratory Rept. 67-1, University of California, June, 1966.

CHAPTER 20

Quality changes
by conventional engineered systems

INTRODUCTION

For the purpose of this discussion of water quality management a
distinction is made between natural and artificial environments in which
changes of quality may be brought about. Ponds, lakes, flowing streams,
estuarine and marine waters, and the soil mantle of the earth are
considered as natural environments, albeit subject to exploitation and
optimizing by engineering works. In contrast, artificial environments are
considered to be those maintained by suitable operational and scientific
techniques within structures designed and constructed by man. A
conventional engineered system, then, is one in which a series of units,
each designed to exploit some particular process or phenomenon, are
put together in an optimum manner to accomplish a desired change in
water quality. Specifically, the engineered works designed for filtration,
sedimentation, aeration, softening or deionizing, and sterilizing water
or for fermenting, oxidizing, precipitating, or otherwise altering or removing
any of its acquired constituents, are herein treated as conventional systems.

The processes involved in conventional systems are those which occur
in nature under appropriate environmental conditions. The objective
of the conventional system, however, is to provide structures in
which water may be isolated from the water resource while these processes
of quality change are accomplished under conditions that bring them
about most rapidly and effectively. In achieving this objective new and
complicating factors may be introduced. For example, the flowing
stream receiving a burden of organic wastes, described in Chapter 18,
degrades and recovers in quality in a sequence of environments. Ideally,
if the streamflow and temperature relationships remained constant,
the waste products of one group of organisms would be flushed away by
flowing water to become the food supply of the next group downstream.

Obviously such an ideal situation does not occur for many reasons, ranging from flow variation to the fact that organisms themselves may be at the mercy of the current. Nevertheless, however dangerously a species may be forced to live, nature establishes an equilibrium such as depicted in Figure 18-1, in which a succession of organisms predominate. In contrast with this system the conventional system of sewage treatment must seek to carry out all these processes within limited space. In the anaerobic digester, for example, the succession of organisms involved in biodegradation (Figure 2-3) must all function simultaneously within a single tank or series of two tanks. Thus environment is folded on environment and the welfare of each group of organisms depends upon the ability of the next to keep its waste products reduced to tolerable levels. Thus there is never a time in the life of an anaerobic digester when any of its several populations may falter either in numbers or efficiency if catastrophy is not to result. For the engineer this generates the familiar difficulties of digester operation and a constant threat of imbalances that man is as yet little prepared to correct.

In a similar fashion, the aerobic cycle of degradation in an activated sludge tank depends upon the harmonious functioning of a mass culture of organisms. Here the watchful eye of experienced chemists and biologists is required to accomplish the results readily achieved by unattended nature, albeit at the expense of the quality needs of other users of the water resource.

The nature of conventional systems is well known to engineers and is described in many textbooks and published articles. Consequently these systems are not presented here in the same detail as are soil, pond, and estuarine systems. Instead the processes exploited by conventional systems are summarized and evaluated in respect to their ability to change the quality of water at the present level of technology and economic acceptability. In this manner any gaps between the public objectives of water quality management and man's ability to attain such objectives can be identified and evaluated.

SUMMARY OF
CONVENTIONAL PROCESSES

Many of the processes utilized in conventional systems of water quality control are applicable both to preparing water for some specific beneficial use or range of uses and to upgrading the quality of return flows to levels required by resource quality objectives. Others are predominately used in one or the other of these two situations. In general, at the present time, those processes designed to stabilize degradable organic matter are applied primarily to return waters, whereas demineralization, when it is used at all, most commonly precedes beneficial use.

The normal applications of each type of process are indicated in Table 20-1. While the table by no means covers all that has been learned concerning the conditions under which each process is optimized, it does indicate how extensively the knowledge of process performance is related to treatment systems as a whole rather than to the individual processes which make up a system.

From the standpoint of water quality management, a consideration of conventional engineered systems underscores the fact that such systems are heavily oriented to the objectives of control of oxygen demand and of coliform organisms in receiving waters. These have been problems of first-order magnitude in the past and will continue to be so in the future. Nevertheless, in looking back over the history of water quality control by engineered systems, one cannot but be impressed at the little change in process that has occurred within the past 40 years. This period has seen the rise of industry as a major producer of wastes, as well as the growth of population which has overwhelmed the resource concepts suited to a pioneer society. In the waste-water treatment field this period has likewise witnessed a thoroughgoing exploration of the limits of conventional approaches. Coagulation with lime or alum, and application to a trickling filter, have characterized the response to each new waste problem, be it one of toxic metals or synthetic organic compounds. Thus it has been well established that new approaches to water quality management must be forthcoming, either at the production or at the treatment end of the scale, if exotic organics, mineralized waters, growth factors in effluents, and similar substances are to be dealt with as effectively as are the biodegradable fractions of organic residues. While effective use of the capacity of soil and pond systems is a necessary adjunct, it will not be enough to offset the trend of public policy toward limited acceptance of man-generated quality factors in the water resource. This means that the engineered system of a "conventional" type must grow in importance. It must also grow in sophistication.

Obviously, an engineered system must be designed to exploit some known phenomenon. The future need, however, is for reduction in the mineral content of return flows, and in some cases, of the available water resource itself. Interestingly, and fortunately, the processes of demineralization listed in the table are those which have been least exploited and are currently under greatest experimental study. It seems safe to predict at this time that the technology of conventional engineered systems will in the future expand in this direction under a revised concept of economic feasibility.

TABLE 20-1 *Summary of conventional processes and systems*

Type or process	Common application	Approximate limit of quality input	Principal change in quality factors (approximate)	Reference
		Gravity separation		
	Reduction in suspended solids in raw water to be pumped	No theoretical limit: 3000–5000 mg/liter typical maximum in floodwaters	Removes larger and heavier suspended solids	
	Primary sewage treatment	Unspecified	50% reduction in suspended solids 35–40% reduction in BOD 50% reduction in turbidity	
	Secondary sewage treatment	Unspecified	Unreported	
	Concentrating return activated sludge (secondary treatment)	Unspecified	Thickens sludge to 20–25% original volume	
Plain sedimentation	Concentrating or reducing suspended solids in industrial wastes, organic and inorganic	Unspecified	Highly dependent upon nature of waste treated	
	Grit removal—raw sewage	Unspecified	Removes heavy suspended solids not transported at velocity of 1 ft/sec	

Plain sedimentation plus skimming	Primary sewage treatment	Unspecified	25–40% reduction in BOD 40–70% reduction in suspended solids 25–75% reduction in bacteria 2% reduction in detergents	[1]
	Various industrial wastes	Unspecified	Dependent upon nature of waste	
Trickling filter plus plain sedimentation	Secondary sewage treatment	0.25–3.0 lb BOD/cu yd/filter	80–95% reduction in BOD 70–92% reduction in suspended solids 90–95% reduction in bacteria 30–35% reduction in ABS 80–90% reduction in LAS	[1] [1]
	Organic industrial wastes (e.g., milk process)	Dependent upon waste treated	Dependent upon nature of waste	
Activated sludge plus plain sedimentation	Secondary sewage treatment	Unspecified	80–95% reduction in BOD 85–95% reduction in suspended solids 95–98% reduction in bacteria 50% reduction in ABS 90–99% reduction in LAS	[1] [1]

TABLE 20-1 *Summary of conventional processes and systems (Continued)*

Type or process	Common application	Approximate limit of quality input	Principal change in quality factors (approximate)	Reference
Gravity separation (Continued)				
Sedimentation after mechanical flocculation	Raw sewage (experimentally)	Unspecified	64% reduction in turbidity 40% reduction in suspended solids 60% reduction in BOD	
	Industrial wastes	Unspecified	Variable, depending upon nature of wastes treated	
	Municipal and industrial water supply Water softening	Unspecified	Seldom evaluated separate from filtration	
Sedimentation after chemical coagulation	Raw sewage (not common)	Unspecified	50–85% reduction in BOD 70–90% reduction in suspended solids 40–80% reduction in bacteria	
	Industrial wastes	Dependent upon waste	Variable, dependent upon nature of waste	

Process	Application		Results	
Chemical coagulation plus sedimentation	Municipal water supply	Unspecified	Coalesces and precipitates dispersed clay colloids Reduces turbidity Reduces color	
	Phosphate removal from waste waters	Unspecified	Reduces soluble phosphates to trace amounts	
	Lime-soda softening of water supplies	Applicable to waters containing Ca and Mg sulfates and bicarbonates Iron and Mg in natural waters (e.g., maximum from 10 mg/liter; minimum, 3 mg/liter)	Reduces hardness to approximately 75 mg/liter; by excess lime to 30–50 mg/liter; by hot process to < 10 mg/liter as $CaCO_3$ Reduces Fe to 0.1 mg/liter ($\pm$) Removes CO_2 – requiring restabilization 80–100% reduction in bacteria by excess lime	[2]
Filtration [2, 3, 4]				
	Tertiary treatment of sewage effluent Water reclamation systems	Relatively low turbidity	90–95% reduction in BOD 85–95% reduction in suspended solids 95–98% reduction in bacteria 90–99% reduction in surfactants	
Slow sand (gravity)	Municipal water supply	Turbidity 40 mg/liter	99% reduction in bacteria 95–100% reduction in turbidity 30% reduction in color Odors and tastes removed 60% reduction in iron	
	Industrial wastes	Unspecified	Varies with nature of waste	

TABLE 20-1 *Summary of conventional processes and systems (Continued)*

Type or process	Common application	Approximate limit of quality input	Principal change in quality factors (approximate)	Reference
Filtration [2, 3, 4] (Continued)				
Rapid sand (gravity)	Municipal and industrial water supply (little used without coagulation)	Low turbidity, e.g., 50 mg/liter, maximum coliform MPN 5000/100 ml	95% reduction in bacteria 90% reduction in turbidity	
Rapid sand plus chemical coagulation (gravity)	Municipal and industrial water supply	No limit specified for maximum turbidity Maximum coliform MPN 5,000–20,000/100 ml	90–99% reduction in bacteria 100% (−) reduction in turbidity Color reduction to less than 5 mg/liter Alkali increased 7.7 mg/liter/gr. alum CO_2 increased 6.8 mg/liter/gr. alum Slight reduction in iron Odor and taste partially removed	
Rapid sand plus chemical coagulation, chlorination, and activated carbon	Municipal and industrial water supply	No limit specified for maximum turbidity Maximum coliform MPN 5,000–20,000/100 ml	Approximately 100% reduction in bacteria 100% reduction in turbidity Color reduced to near zero Iron and Mn reduced Taste and odor removed	
Rapid sand (pressure) (precoat with chemical floc)	Small municipal supply Swimming pools Industrial supply and process Emergency and military use	Generally unspecified Low turbidity desirable	Similar to rapid sand filter but more variable in performance	

Process	Applications	Requirements	Performance	Reference
Diatomaceous earth (pressure and vacuum)	Small municipal supplies Institutional water supply Swimming pools Industrial supply and process Emergency and military use	None specified, but operation depends upon nature of water	Capable of good clarification of water; efficiency, however, not well documented 40–90% reduction in suspended solids 50% reduction in color	[5]
Contact filters	Manganese removal Iron removal	None specified	Reduces to USPHS Standards 88% reduction in iron	
Bag filters	Swimming pools	Unspecified	Strains out hair and coarser suspended solids, reduces bacteria to level controllable by chlorination practice	
Microstraining	Primary clarification of water prior to filtration Clarification of sewage effluents Treatment of industrial wastes	Size of particles to be removed greater than screen size Material suitable for microstraining	87–96% reduction in microscopic organisms 60–90% reduction in microscopic particulates 50–60% reduction in suspended solids trickling filter effluent 30–40% reduction in turbidity	[7] [8] [9] [7]
Fine screening	Raw sewage	None specified	5–10% reduction in BOD 2–20% reduction in suspended solids 10–20% reduction in bacteria	
	Industrial wastes (e.g., cannery, pulp mill, etc.)	None specified	Varies with nature of waste	
Carbon filters	Special municipal and industrial water applications	Very low turbidity, other not specified	Adsorbs exotic organic chemicals, including surfactants Removes tastes and odors Adsorbs miscellaneous gases	

TABLE 20-1 *Summary of conventional processes and systems (Continued)*

Aeration

Type or process	Common application	Approximate limit of quality input	Principal change in quality factors (approximate)	Reference
Spray or cascade	Municipal and industrial water supply Industrial waste treatment	Unspecified	Releases gases producing taste and odor Reduces CO_2 in groundwaters to normal surface water levels Partial removal of H_2S Partial removal of gases of decomposition Oxidation and removal of soluble iron in groundwaters; 80–97% reduction observed	[5]
Pressure aerators	Treatment of sewage and industrial wastes	Limits variable or unspecified	Grit precipitated Grease concentrated at surface Separates various solids by flotation Maintains aerobic conditions in biological systems, e.g., activated sludge, aerated ponds Reduces ABS or LAS 1–2 mg/liter Reduces septicity of sewage	
Oxidation ponds (see Chapter 17)	Treatment of domestic sewage and organic industrial wastes	No toxic substances (see Chapter 17)	75–96% reduction in BOD 90–99% reduction in suspended solids 98–99.9% reduction in bacteria 56–93% reduction in LAS	

Demineralization

Ion exchange (natural or synthetic zeolite)	Softening of groundwater supplies for municipal or industrial use	Hardness (Ca and Mg sulfates and bicarbonates) of natural waters $> 850–1000$ mg/liter $CaCo_3$, Iron $< 1.5–2$ mg/liter Low in silica $CO_2 < 15$ mg/liter	Increases sodium content by exchange with removed Ca and Mg	[2]
Ion exchange (greensand or styrene base gels)	Iron or Mn removal from groundwater	Iron less than approximately 2.0 mg/liter	90-100% removal of iron Mn partially removed	[5]
Ion exchange (organic cation exchangers)	Special water conditioning for industry and commerce	Unspecified	Removes all cations (Na, K, Mg, Fe, Cu, Mn)	
Ion exchange (anion exchangers)	Special water conditioning for industry and commerce	Unspecified	Removes SO_4, Cl, NO_3, etc.	
Ion exchange (fluoride exchangers)	Defluoridation of public water supply	More than 1.5 mg/liter F in water supply	Approximately 100% removal possible Normally reduced to < 1.5 mg/liter	

TABLE 20-1 *Summary of conventional processes and systems (Continued)*

Demineralization (Continued)

Type or process	Common application	Approximate limit of quality input	Principal change in quality factors (approximate)	Reference
Electrochemical desalting	Reclaiming water from saline sources, public and industrial supplies Demineralizing municipal waste effluents	Applicable to highly saline or brackish waters	Removes anions and cations	[10]
Reverse osmosis	Reclamation of water from brackish natural or waste waters (experimental)	Brackish waters, upper limit not specified	Reduces ions depending upon concentration difference across membrane 97–98% reduction in TDS, ABS, and COD	[10]
Distillation	Reclamation of water from saline sources Specialty industrial and commercial supplies	No limit	Produces distilled water (may be contaminated with NH_3, volatile organics, etc.)	
Freezing	Reclamation of water from saline sources Specialty industrial and commercial supplies	No limit	(Experimental)	

Chlorination

Liquid Cl$_2$ and Cl$_2$ compounds	Public water supply Industrial water supply	Turbidity low for waters to be sterilized by Cl$_2$	Reduces bacterial load on filters Oxidizes organic matter Reduces odor Assists in color removal 100% (−) reduction in bacteria Controls plankton growth in reservoirs Reduces Mn concentration at breakpoint
	Municipal and industrial waste-water treatment and management	Unspecified	Assists in grease removal Controls filter fly nuisance Cleans air stones in aeration systems Removes H$_2$S Removes NH$_3$ Controls slime formation in sewers and cooling towers Assists in control of digester foaming Disinfects effluent; 98–99% reduction in bacteria

Digestion

Anaerobic digestion	Stabilization of sewage solids Stabilization of organic industrial wastes	pH above 6.8 Acids limited No toxic substances in significant amounts Minimum of grit	Reduces organic sludges to humus and relatively stable chemical compounds Produces offensive supernatant

REFERENCES

1 S. A. Klein and P. H. McGauhey: "Degradation of Biologically Soft Detergents of Wastewater Treatment Processes," *J. Water Pollution Control Federation*, 37(6), June, 1965.

2 G. M. Fair and J. C. Geyer: *Elements of Water Supply and Waste Water Disposal*, John Wiley & Sons, Inc., New York, 1957.

3 J. W. Clark and W. Viessman, Jr.: *Water Supply and Pollution Control*. International Textbook Company, Scranton, Pa., 1965.

4 *Water Quality and Treatment*, American Water Works Association, New York, 2d ed., 1950.

5 K. E. Cowser: "Iron Removal Practices with Illinois Ground Waters," *Water and Sewage Works*, 98(12), 1951.

6 G. J. Turre: "60-mgd Microstraining Plant Meets Denver's Growing Needs," *Water Works Eng.*, December, 1961.

7 Rolf C. Carter et al.: "Behavior and Evaluation of Microstraining for the Lytle Creek Supply," Paper presented at California Section of American Water Works Association, Sacramento, October 27, 1961.

8 C. E. Keefer: "Operating a Sewage Microstrainer," *Water and Sewage Works*, 101(7), 1954.

9 "Summary Report, Advanced Waste Treatment Research," January 1962–June 1964, AWTR-14, U.S. Public Health Service, April, 1965.

CHAPTER 21

Summary of the status
of water quality management

INTRODUCTION

The concept that management of the quality of water is quite as important as its physical management is now widely recognized and advocated in the United States. However, it is of such recent origin that no precise definition of the term has been established and widely accepted, principally because the jurisdiction in which the responsibility for management falls is exceedingly fragmented. Local, state, and Federal agencies, often institutionalized within limited or narrow frameworks, are responsible for a variety of aspects of water quality control—sometimes conflicting, sometimes overlapping. Moreover, the legislation which established these agencies was necessarily based on the rationale of the time which, as noted in previous chapters, was not one of broad concern for the overall resource.

THE RISE OF A CONCEPT
OF QUALITY MANAGEMENT

As discussed in Chapter 3, three conceptual stages are identifiable along the road to quality management as a major aspect of resources conservation. Each represents a distinct extension of the objectives of society. Growing out of the disasters of less sanitary times, protection of the public health was the first concern for impurities in public water supplies. Pollution with human or animal wastes was the factor to be prevented by watershed management or overcome by water purification techniques. Here pollution was defined first in terms of disease potential and later in terms also of offenses to the senses of sight and smell. Significantly, "pollution" was the crime and "pollution control" the punishment; and the task of the control agency was to make the punishment fit the crime.

Protection of multiple beneficial uses became an added objective of society about midcentury. A need to take cognizance of yet another situation requiring water quality control arose from several factors, such as:

1. Growth of population and its concentration in urban centers
2. Advance in per capita use of water
3. Emergence of industry as a major producer of wastes
4. Lengthening of the spectrum of wastes far beyond organic residues of life processes
5. The growing problem, in some regions, of mineralized agricultural return waters

Public response to this need was the addition of "water pollution control" agencies to those existing to protect the public health. In some states the new duties were assigned to the department of public health, hence likely to become constrained within traditional concepts of pollution control. In other states, new agencies were created, to coexist with the health department in an uneasy balance of authority. The word "pollution" in the title of these agencies underscores the fact that changed conditions, specifically an expansion of the goals of society from health protection to protection of the interests of other beneficial uses as well, were not immediately reflected in the rationale underlying our concepts of the problem created by waste residues in the freshwater resource. Impurities were "pollutants"; the producers were "polluters"; and the remedy remained as "pollution control."

Although under this concept attention was directed to the nature and control of pollutants, pollution became almost impossible to define because the quality needs of one beneficial use were so different from those of another that no overall common concept was possible. Thus it became necessary first to say which beneficial uses were to be protected before pollution could be said to occur at all. And thus a concept of quality began to replace the traditional concept of pollution.

It remained for the third and most recent extension of public objectives to make necessary a concept of water quality as the aspect to be controlled. This concept emerged from such factors as the urban-industrial economy of plenty which gave rise to both a vast need for recreation and the leisure time in which to pursue it. It took the form of such social or aesthetic goals as clean water per se, maintenance of wildlife habitat, preservation of wilderness areas, quality of the environment, and even quality of life itself. However unrelated or unrealistic such goals may seem in many instances, they relate to the nature of the resource itself rather than to the agent (or pollutant) which depreciates that nature. Therefore, it became evident that if protection of the resource was to be the objective, the concept had to be broad enough to permit the management of water resources in an effective manner for a variety of social objectives. Without such a concept, the residues of beneficial use of the nation's water resource would eventually seriously and disastrously constrain its freedom to use that resource to the benefit of man.

It is out of this growing necessity that the concept of water quality management, to which this work is directed, is emerging, with the overall systems concept as its major rationale, and engineered subsystems as the control devices at critical interchange points.

STATUS OF WATER QUALITY MANAGEMENT

In preceding chapters attention has been directed to the nature of the system to be managed, to the hinges in the system where management techniques may be applied, and to the current status of technology for quality control. In all these aspects of water quality management there are problems which are as yet imperfectly resolved and which impose serious limitations on our ability to manage water quality. Some of the most critical discussed in previous chapters are worthy of recapitulation here.

Of first importance is the problem of defining quality in numerical terms. This dilemma is largely resolved by recognizing that quality per se has no particular meaning except in relation to the needs of individual beneficial uses. From a water resource viewpoint, however, quality may generally be defined in terms of suitability of the water resource for those beneficial uses which do not require withdrawals, and for other uses after treatment by conventional technology. To this end the concentration of various quality factors may be measured quantitatively and properly cataloged. Limitations on this approach are nevertheless serious, because:

1. The range of quality factors traditionally observed is related to too narrow a spectrum—principally those pertinent to the organic cycle of decay and to earth minerals.
2. The resolving power of many analytic methods is too low to detect various exotic refractory compounds.
3. It is difficult to evaluate the significance of low concentrations of various factors which may appear as a result of refinement of analytical techniques, or of some new waste discharge.
4. The effects, particularly long-term effects, of some quality factors are unknown.
5. The synergistic effects of all manner of quality factors on aquatic life are unknown.

Under these circumstances it becomes difficult to find or to assess the significance of many quality factors emanating from industrial and commercial activities. Thus the quality objectives of management are obscure and the selection of processes to be applied through engineered systems becomes difficult or impossible, if indeed appropriate processes have been developed.

In the case of return flows from domestic use, the quality factors interchanged are quite well understood, as are many of the principles which can be harnessed to quality management goals. These processes, however, are directed to the stabilization of degradable organic solids; hence complete perfection of the processes involved still leaves nutrients in the effluent, together with an increase in other mineral constituents. Thus as the population increases and domestic return flows represent a greater percentage of the freshwater resource, a greater degree of purification may be required, involving processes considered today to be technologically problematical or economically infeasible.

The same may be said for industrial return flows. In addition to refractory factors of unknown implications in terms of water quality, a variety of ions and minerals stemming from either process waters or cooling water will add to the mineral load which quality management must consider. Once again the engineer is confronted with a problem of technological and economic feasibility at the current level of process development.

In some areas, particularly where water shortages make resource considerations the most acute, irrigation return waters pose a problem of mineralization that has not been attacked by engineered systems. In combination with industrial return flows and evaporation from surface waters, this leads to an increasing salinity of the freshwater resource with which water quality management must one day become concerned. Despite current concepts of the economics of desalination and deionization, management techniques of the future must certainly involve demineralization. The first steps in this direction are likely to involve removal of phosphates and nitrates from domestic return flows in situations where eutrophication of receiving waters is a major concern. Experiments with chemical processes have been conducted in this area, with one of the most likely being the conversion to algal cells and the subsequent harvesting of the algae.

In the case of irrigation return waters, however, the economy of agriculture, together with conflicting objectives of public policy, will call for a management approach unlike that imposed upon cities and industries.

Of the many environments in which the effect of quality factors is imperfectly understood none is more problematical than estuarine waters. From the standpoint of water quality management, however, the estuarine relationships are so complex and subtle that greater understanding of them is certain to increase concern for the quality of such waters. Thus added constraints may be placed upon freshwater management techniques in the interests of estuarine water quality. The same prospect appears in relation to marine disposal of wastes, where current technology is inadequate to make design possible on the basis of all water quality objectives which may well be imposed.

An unresolved water quality management problem of concern is that of the agricultural need for pesticides versus the effect of residues in the aquatic environment. A great deal more research on time-degradable compounds and on the long-term effects of pesticides in various ecological chains is needed before this problem can be resolved.

Reconciling the inevitability of return flows from beneficial use mingling with the "natural" waters, with the quality changes which are feasible under existing or foreseeable technology and economic attitudes is one of the major factors in water quality management today. Some newer approaches to refined treatment by soil systems and ponds are being developed. Nevertheless, conventional systems have limitations which leave an increasing gap between objectives and their attainability. This is particularly true because water treatment technology has advanced but slowly under the constraints of scientific discovery and the conservatism inherent in public works. Meanwhile the leap from tolerance of polluted water to a demand for wilderness quality in the water resource has been quickly made on the basis of social goals, emotions, and political concern.

Compounding the problem of water quality management still further is the fact that existing approaches to quality control have not yet had time to become freed of the concept of pollution control or of legislation directed to that narrower goal. Objectives of design expressed as criteria or standards that depend upon the local sector of the resource and engineered works to achieve these are likewise directed to the immediate local problem. Thus the overall system of water resources quality management into which these subsystems should fit has not yet been established through public planning and policy decisions.

CONCLUSION

The foregoing problems arising from the inability of institutions and technology to respond quickly to changed conditions represent dislocations which result from a change from use-oriented goals to resource-oriented objectives of water quality control. Nevertheless, the concept of water quality management clearly seems to represent emerging public policy in relation to resources management and conservation. The result is that the spectrum of quality factors which must be managed by engineered systems in the years immediately ahead is so lengthened as to challenge as never before the imagination and ingenuity of engineers, as well as that of the institutions which prepare men for practice in the engineering profession.